If I Understood You,
Would I Have This Look on My Face?

If I Understood You,
Would I Have
This Look
on My Face?

*My Adventures in
the Art and Science
of Relating and
Communicating*

ALAN ALDA

RANDOM HOUSE  NEW YORK

Published in the United States by Random House,
an imprint and division of
Penguin Random House LLC, New York.

RANDOM HOUSE and the HOUSE colophon are
registered trademarks of Penguin Random House LLC.

Grateful acknowledgment is made to Alfred Publishing LLC and
Bienstock Music Publishing Company on behalf of Redwood Music Ltd
c/o Carlin America, Inc., for permission to reprint an excerpt from
"As Time Goes By" (from *Casablanca*), words and music by Herman Huffeld,
copyright © 1931 (renewed) WB Music Corp. All rights reserved.
Used by permission of Alfred Publishing LLC and Bienstock Music
Publishing Company on behalf of Redwood Music Ltd.

LIBRARY OF CONGRESS CATALOGING-IN-PUBLICATION DATA
NAMES: Alda, Alan, author.
TITLE: If I understood you, would I have this look on my face? : my adventures
in the art and science of relating and communicating / Alan Alda.
DESCRIPTION: New York : Random House, 2017.
IDENTIFIERS: LCCN 2016045922 | ISBN 9780812989144 | ISBN 9780812989168 (ebook)
SUBJECTS: LCSH: Interpersonal communication. | Interpersonal relations.
CLASSIFICATION: LCC BF637.C45 A424 2017 | DDC 153.6—dc23 LC record
available at https://lccn.loc.gov/2016045922

Printed in the United States of America on acid-free paper

randomhousebooks.com

246897531

FIRST EDITION

Frontispiece illustration by Barry Blitt

Book design by Carole Lowenstein

for Arlene
my love and my pal

"The single biggest problem in communication
is the illusion that it has taken place."

—often attributed to
GEORGE BERNARD SHAW.
Although it's doubtful
he ever said it.

CONTENTS

CONTENTS

INTRODUCTION

When it finally became clear to me that I often didn't understand what people were telling me, I was on the road to somewhere good.

Some of the things they were trying to communicate were complicated, but that didn't seem like a good reason why I didn't understand them. If *they* could understand these things, why couldn't I? An accountant would tell me about the tax code in a way that made no sense. A salesman would explain an insurance policy that didn't seem to have a basis in reality. It wasn't any consolation when I came to realize that pretty much everybody misunderstands everybody else. Maybe not all the time, and not totally, but just enough to seriously mess things up.

People are dying because we can't communicate in ways that allow us to understand one another.

That sounds like an exaggeration, but I don't think it is.

When patients can't relate to their doctors and don't follow their orders, when engineers can't convince a town that the dam could break, when a parent can't win the trust of a child enough to warn her off a lethal drug, they can all be headed for a serious ending.

This book is about what we can do about that; about how I learned what I believe is the essential key to good communication, and to relating to one another in a more powerful way. Surprisingly, I found that key in my training and experience as an actor, and it's helped me teach others how to communicate better, especially about things that are difficult to talk about or hard to grasp.

IT ALL BEGAN WITH MY TOOTH

The dentist had the sharp end of the blade inches from my face.

It was only then that he chose to tell me what he was seconds away from doing to my mouth. "There will be some tethering," he said.

I froze. *Tethering?* My mind was racing. *What does he mean?* How could the word *tethering* apply to my mouth? He seemed impatient and I didn't want to annoy him, but he was, after all, about to put a scalpel in my mouth. I asked him what he meant by *tethering.* He looked surprised, as if I should know the meaning of a simple word. He began barking at me. "Tethering, tethering!" he said.

I was well over the age of fifty, and certainly old enough to ask him to put the knife down and answer a few questions. But there he was in his priestly surgeon's gown, and getting increasingly impatient. "Okay," I said, a little too accommodatingly. Then he put the scalpel into my mouth and cut.

I didn't realize it at the time, but this was a watershed moment for me.

For the rest of my life, I would be living with the results of these few seconds of poor communication—in ways that were both good and bad.

First the bad: A few weeks later, I was acting in a movie. The camera was in close on me as I smiled. A relaxed, happy smile. After the take, the director of photography came over, looking puzzled, and said, "Why were you sneering? I thought you were supposed to smile."

"I *was* smiling," I said.

"Nooo. Sneering," he said.

I looked in a mirror and smiled. I was sneering.

My upper lip drooped lazily over my teeth, and no matter how I tried I couldn't form an actual smile. The problem was my frenum. I no longer had one.

In case you're not familiar with your frenum, it's just above your front teeth, between the gum and the inside of the upper lip. If you put your tongue at the top of your gums above your front teeth, you'll feel a slim bit of connective tissue, or at least you will if a barking dentist hasn't been in your mouth. The tissue is called the maxillary labial frenum, and he had severed mine.

The procedure was one he had invented. It enabled him to pull a flap of gum tissue down over the socket of the front tooth he had extracted. The idea was to give the socket a fresh blood supply while it healed. He was proud of his invention and it seemed perfectly suitable for the gum's blood supply, but not so good for using my face in movies. Without my frenum, my upper lip just hung there like a scalloped drape in the window of an old hotel.

After the movie shoot, I called him and explained with saintly patience that I made a living with my face and sometimes I needed one that could smile.

His response was curt. "I told you there were two steps to the procedure. I haven't done the second step yet." I was a little reluctant to let him do the second step. Maybe this time he'd have a go at the frenum under my tongue. I didn't have many frena left, and he seemed to have an unnatural attraction to them.

A couple of weeks later, I got a letter from him that was formal and cold. No hint that he was anything like sorry that I was feeling a little mutilated. It was clear that the point of the letter was to lay out his defense and discourage a possible lawsuit. Until I saw the tone of his letter, I hadn't even thought of suing (and I never did)—but if he wanted to avoid a suit, he was going about it in exactly the wrong way.

The experience wasn't all bad, though. For one thing, I learned to work around my frenum-less smile, and my new, slightly off-kilter grin enabled me to play a whole new set of villains. Even better, that moment in the dentist's chair was useful in ways I wouldn't understand at the time.

I've come to see my exchange with the dentist that day as something that happens frequently in life—a brief encounter that threatens a relationship's delicate tissue, the tender frenum of friendship. I wasn't looking for friendship that day, but at least I wanted the feeling that I was actually being seen by him. Even though his gaze was intense, I realized that as far as he was concerned I wasn't really there—not as a person. If I was there at all, I was something on his checklist. He was speaking into the vague mist of interpersonal nothingness.

Those few minutes I spent in his chair have become a symbol for me of really, really poor communication and of what causes it: disengagement from the person we hope will understand us. That disengagement can stand in the way of all kinds of happiness and success, from the world of business to the business of love.

Not being truly engaged with the people we're trying to communicate with, and then suffering the snags of misunderstanding, is the grit in the gears of daily life.

It jams our relations with others when people don't "get it," when they don't understand what we think is the simplest of statements.

You run a company and you think you are relating to your customers and employees, and that they understand what you're saying, but they don't, and both customers and employees are leaving you. You're a scientist who can't get funded because the people with the money just can't figure out what you're telling them. You're a doctor who reacts to a needy patient with annoyance; or you love someone who finds *you* annoying, because they just don't get what you're trying to say.

But it doesn't have to be that way.

For the last twenty years, I've been trying to understand why communicating seems so hard—especially when we're trying to communicate something weighty and complicated. I started with how scientists explain their work to the public: I helped found the Center for Communicating Science at Stony Brook University in New York, and we've spread what we learned to universities and medical schools across the country and overseas.

But as we helped scientists be clear to the rest of us, I realized we were teaching something so fundamental to communication that it affects not just how scientists communicate, but the way *all* of us relate to one another.

We were developing empathy and the ability to be aware of what was happening in the mind of another person.

This, we realized, is the key, the fundamental ingredient without which real communication can't happen. Developing empathy and learning to recognize what the other person is thinking are both essential to good communication, and are what this book is about.

This is a personal story, too. It's about what I've learned over the years as an actor that can help us all be clearer with one another, including about complicated things. Some of what I've learned has come from talking to smart people about their research, but much of it stems from the experience of standing face-to-face on stage with another actor. It's changed the way I engage with other people in my daily life. And it can be learned by anyone, not just by those touched with a talent for acting. It's an amazingly simple thing, a power we're built to use, and yet too often ignore.

In acting, we call it relating.

PART ONE

Relating Is Everything

CHAPTER 1

Relating: It's the Cake

A couple of decades ago, a letter came in the mail that set me on a path that would not only bring me to a deeper understanding of that day with the dentist, but would actually change the direction of my life.

The letter was from a television producer, asking if I would be interested in hosting a show on PBS called *Scientific American Frontiers*. I was in love with science and had read every issue of the magazine *Scientific American* since I was a young man. It had been my only education in the subject. I was so excited I had to read the invitation twice. *Scientific American!* My alma mater! This would be a chance to actually learn from scientists themselves.

After a few minutes, though, I realized that the producers were probably only looking for someone well known to appear at the beginning of the show to introduce that week's topic and then disappear to read an off-camera narration. That sounded like a lot less fun than talking to scientists, so I asked them if, instead, I could interview the scientists on camera. I knew if we'd be shooting interviews I'd spend the whole day with them—not just on camera but during the hours of setting up, having lunch, and wandering around their labs. I'd have a chance to learn something.

There was one minor hitch. I didn't have much experience interviewing anybody, let alone scientists, so if the producers agreed, they'd be taking something of a chance on me.

I was, however, blissfully confident. I had taken over as guest host on talk shows a few times, but more than that, I thought that one of the tools of my profession ought to help: the ability to listen and react. Plus, I had been trained in improvisation, a particular kind of theater training—games and exercises that enable you to open up to another person, to tune in to them, to engage with them in a dance of ideas and feelings, and to go anywhere it takes you, together.

I'm sure the producers of *Scientific American Frontiers* weren't as confident of all this as I was, but they decided to take a chance on me. We began shooting the series in 1993.

The first story we shot featured racing cars powered only by solar energy. We went out to the California State University, Los Angeles, and set up in a workshop where a scientist was working on a large solar panel. This would be my first actual science interview. John Angier, one of the producers, called me over and nodded in the direction of the scientist. Peter Hoving, the cameraman, lifted his camera and started rolling.

As I entered the room, I didn't realize that this would be the beginning of more than twenty years of trying to figure out what makes communication work, getting beyond the impoverished ways of the barking dentist, looking for empathy and a deeper kind of listening in almost every part of my life. This moment would begin it all. But not only didn't I realize this, I was vaguely aware that I didn't really know what I was doing. I hesitated for a moment.

John Angier nodded toward the scientist again and, with just a slight air of *Well, this is what you wanted,* he said, "Go on. Go in there and start talking."

I walked over to the scientist, smiled confidently—and immediately made three huge blunders.

LISTENING WITH EYES, EARS, AND FEELINGS

My first blunder was assuming that I knew more than I did.

After a brief hello and a quick glance at his solar panel, I told the scientist how amazing it was that he had put all this together just using parts off the shelf. I saw his face tighten a little. "They're not off the shelf," he said, slightly offended. "We had to *make* a lot of them." I saw the anxiety in his face, but I didn't respond to it. I experienced a little anxiety of my own, but I ignored both his and mine. Instead, I made the next blunder with my body.

I reached out to the solar panel and laid my hand on it, assuming a bit of unearned familiarity. I saw something happen to his face again, but I kept going. Not content with touching the panel, I gave the thing an affectionate pat. "Amazing," I said, hoping that time would pass more comfortably if I showed a measure of awe.

"Please don't touch the panel," he said. "You could ruin it." The distress in his face was now very clear to me. I had seen it earlier, but somehow I had ignored it. I hadn't been listening with my eyes.

I lurched through a couple of questions about solar panels, but the interview was lame and halting. I was well into my third blunder: Just as I hadn't been really relating to him in not responding to the look on his face, none of my responses grew out of what he was telling me. I wasn't really listening to him when he answered my questions.

In fact, I hadn't been listening in three different ways. When I'd told him he had made the panel with parts off the shelf, I was paying more attention to my own assumptions than I was to him. When I didn't read his face, I wasn't listening to his body language. And when my questions didn't spring from what he was saying, I was disconnected from him. I was alone. How could the conversation have been anything but strained when I had shut myself up in my own head?

I was a little downcast by the experience. Where was the improvisational ability to listen and react that I knew how to do onstage, that I had been trained in and was so proud of? I cherished my experiences in improvising with other actors. Why wasn't I doing it now?

IMPROVISING

Improvising on the stage is usually thought of as creating funny sketches on the spot, with no preparation. Most improvisation that audiences are exposed to is comedy improv, and that was my first experience with improvisation.

One summer in my early twenties, I was performing in a cabaret show, sunk in the basement of a hotel in Hyannis Port. The first act of our show was a set of sketches we had created in rehearsal through improvisation. There was no writing of funny lines; it was all developed through the spontaneous interactions of the actors. The only preparation beforehand was thinking about characters we could play, and figuring out the quirks they had that could be relied on no matter what the other actor tossed our way. We worked these sketches over many times in rehearsal, and, although they were derived through improvisation, we knew we had these surefire set pieces for the first half of the show.

The second half was much scarier.

Before the intermission, I would ask the audience to give us words or headlines from the news. Then we'd take this list of minimal prompts backstage, and for fifteen furious minutes we'd toss ideas back and forth.

"They gave us the word *taxes*," I might say to Honey Shepherd (who went on, decades later, to play Carmela's mother on *The Sopranos*). "How about if you do your Nice Old Lady being audited for her tax return?" Her character was sweet, reasonable, and totally antiwar.

When she got out on stage, if the tax auditor asked her why she hadn't paid her taxes, Honey's Nice Old Lady could be relied on to say something like "I don't want to buy a bomb." Which would lead to a tangled, logically illogical dialogue.

As we brainstormed the most minimal of premises for sketches, we had no idea what we would actually do or say. We didn't know where a sketch would go or how it would end. Whoever was offstage during a scene would have their hand on the light switch, and when something funny happened that sounded like a concluding moment, they'd flip the switch and we'd have the punctuation of a blackout.

In an improvised press conference, I would take on the persona of President John Kennedy, answering questions from journalists in the audience who had asked the same questions of the real John Kennedy in the same hotel that morning. His answers to their questions hadn't made it into the newspapers yet, so I was flying completely blind.

It was easy to worry that we would fail and flame out in front of the audience.

As opening night approached, we felt a thrill that must be like the thrill that runs through a person's tingling body just before he jumps off a bridge.

As scary as this kind of improvising was, there was excitement in not knowing what we would suddenly be doing during a show. It was exhilarating, but we were limited by two things: We had to be funny, and we had little or no training in improv. We were relying mainly on sheer guts.

A year or two later, though, I was introduced to a completely different form of improvising.

I was invited to join a workshop conducted by Paul Sills, who had founded Second City, the phenomenally successful improv company. We met twice a week on the Second City stage in downtown New York City. It was the same stage where skilled comedy improvisers would perform nightly, but in these sessions Paul introduced us to a completely different kind of work.

His mother, Viola Spolin, had done groundbreaking work in creating a kind of improvisation training that was rigorous and exacting, and that slowly built in actors the ability to connect with one another spontaneously. Comedy was not the objective. Cleverness and joke making were forbidden. Something else, something much more fundamental to theater, was being explored: a kind of relating that could lead to deeper, more affecting performances.

At each session, Paul would open Viola Spolin's book, *Improvisation for the Theater,* and lead us through games and exercises that little by little transformed us. The games connected each of us to the other players in a dynamic way. What one player did was immediately sensed and responded to by the other player. And that, in turn, created a spontaneous response in the first player. It was true relating and responsive listening, which, I've come to realize, is necessary on the stage and in life as well.

After six months, I felt that these improv sessions had changed me both as an actor and as a person.

But here I was now in this interview about solar panels, and it wasn't working. I was talking to a scientist who could give me the knowledge I craved—and *I wasn't listening.*

Why? I had spent my whole life on the stage listening to the other actors. Or trying to. But it seemed to be something I constantly had to relearn.

When I started out as an actor, I had the vague awareness that listening had something to do with relating to the other person, although *relating* was a word with a mysterious ring to it. I heard it often from directors and had seen it repeatedly in books by the Russian acting gurus—Stanislavski, Boleslavski, and one or two other avskis. But I was still hazy about what relating actually entailed. It obviously had to do with making some kind of contact with another person. I drew the natural conclusion that it meant putting myself in the other person's face. So when I was asked to relate more, I would tilt over in their general direction, in the manner of an errant telephone pole. But this wasn't actually relating; it was just leaning over. If the director asked for even more relating, I would slump my shoulders and position my nose even closer to the other actor's. I would be hunched over like the ape in the evolution cartoon, just before he straightens up and walks like a human. It didn't make directors sigh in admiration.

Once, a long time ago, Mike Nichols was directing Barbara Harris and me in a rehearsal for the Broadway musical *The Apple Tree.* He asked us both several times to relate better—although he seemed to be looking more in my direction than in hers. Finally, he couldn't stand it anymore. "You kids think *relating* is the icing on the cake," he said. "It isn't. It's the *cake.*"

So, what is it? What's relating? What's the cake? It took me years to be able to put it roughly into words.

It's being so aware of the other person that, even if you have your back to them, you're observing them. It's letting everything about them affect you; not just their words, but also their tone of voice, their body language, even subtle things like where they're standing in the room or how they occupy a chair. Relating is letting all that seep into you and have an effect on how you respond to the other person.

RESPONSIVE LISTENING

There's a body of scientific literature on responsive listening, but I came to understand it in a personal way through my work. In acting, this kind of relating is fundamental. You don't say your next line simply because it's in the script. You say it because the other person has behaved in a way that *makes* you say it. Relating to them allows them to have an effect on you—to change you, in way. And *that's* why you respond the way you do.

For an actor, it's the difference between planning how you're going to behave, which looks like acting, and finding your performance in the other person's eyes, which makes you respond to one another—and which looks like life.

But, with all that behind me, here I was supposedly in conversation with the solar panel scientist, and I wasn't relating to him. Slowly, it was beginning to dawn on me: It's not just in acting that genuine relating has to take place—real conversation can't happen if listening is just my waiting for you to finish talking.

LISTENING AND WILLING TO BE CHANGED

I so loved this idea—that on the stage the other actor has to be able to affect you if a scene is to take place—that I came to the conclusion that, even in life, unless I'm responding with my whole self—unless, in fact, I'm willing to be *changed* by you—

I'm probably not really listening. But if I do listen—openly, naïvely, and innocently—there's a chance, possibly the only chance, that a true dialogue and real communication will take place between us.

This was the first step in understanding what had to take place before doctors (and *dentists*) could talk with their patients; before people in business could relate to their customers, parents could advise their children, and couples could work together—with far fewer misunderstandings and hard feelings. At first, though, I was concentrating on helping scientists get their story out in the most human-sounding way.

Once I began to relearn listening as a human interaction, and not just an acting technique, I could go into each interview for *Scientific American Frontiers* without a set of questions. It wasn't really an interview anymore. It was a conversation.

After a while, I saw that I was having trouble talking with them whenever I thought I knew more than I really did about their work. I was boxing in the scientists with questions that were based on false assumptions. I took a bold step and stopped reading the scientists' research papers before I met with them. I would come in armed only with curiosity and my own natural ignorance. I was learning the value of bringing my ignorance to the surface. The scientists could see exactly how much I already understood, and they could start there.

Ignorance was my ally as long as it was backed up by curiosity. Ignorance without curiosity is not so good, but *with* curiosity it was the clear water through which I could see the coins at the bottom of the fountain.

CONTAGIOUS LISTENING

It led to a dynamic relationship. The scientists were glad to see that I really wanted to understand their work, and, just as it does in improv games, it had an effect on them and they themselves became more responsive. They could relate to me as a person. They stopped worrying about the camera and about the audience on the other side of the lens. They stopped feeling compelled to speak in highly technical language. Their real humanity came out, because they had someone in conversation with them who insisted on understanding them, no matter how long it took.

My responsive listening encouraged theirs. It was contagious. They were drawn into a kind of dance, and suddenly we had life happening between us.

When their tone of voice became warm and intimate and their natural sense of humor came through, the audience was able to see scientists as fellow people, sometimes with very human traits.

On one show, I was talking with a researcher who had created a surprisingly intelligent robot. Some people worry about robots and I asked him how he'd feel if his machines someday got so good at manufacturing themselves that they decided they didn't need humans anymore and totally turned against the human race. He thought about it for a long moment and then asked, "Well, would I win the Nobel Prize?"

I don't think he'd have made that joke in a straight interview, and it was a delightful moment that candidly touched on the competitive urge that drives much of science.

Through all of this, we had stumbled into a way of allowing scientists to emerge as themselves, and allowing the audience to see that the people who emerged were warm and engaging. There was a bond between us that the audience could recognize as a common human interaction.

But then one day I saw how easily that bond can be broken.

I was in the office of a scientist whom I found fascinating, and we were enjoying that special connection. We were sailing. Even as she talked about the complexities of her work, I felt we were on the same wavelength. We were relating.

Then she dived just a little deeper into her work and I saw something flicker across her face. I didn't know what it was, but suddenly it wasn't about us anymore. She had been reminded, I think, that what she was telling me was just like a lecture she usually gave. This must have been what was happening, because she suddenly turned to the camera, looked right into the lens, and started lecturing it. The tone of her voice went from warm and natural and personal to formal and lecture-hall stilted. Her vocabulary became totally incomprehensible. She was in flat-out lecture mode. And the camera was getting the brunt of it.

I coaxed her back with some naïve questions. It wasn't hard to come up with naïve questions, because I really couldn't understand what she was saying. Slowly, she turned back to me, caught my eye, and was warm again. We were sailing again. This glorious feeling lasted for about forty-five seconds, at which point she couldn't stand it anymore, turned back to the camera, raised an imaginary lecture stick, and gave the camera a good thrashing.

This was a rare experience on *Scientific American Frontiers*. Most of the time, once we locked into conversational contact, our rapport would strengthen and grow. And I don't at all blame the scientist for our loss of rapport that day. She was vibrant and engaging and gave us a wonderful segment. But I saw how the pull of formality and jargon can yank someone into not relating.

When the series ended after eleven years on the air, I had the feeling we had accomplished something of value, bringing science to people in a way that was a little more inviting, yet totally accurate. Still, something bothered me. I kept returning in my mind to that scene in the scientist's office. If I hadn't doggedly pulled her back from it, she might have stayed in lecture mode.

I wondered: What if scientists could be helped to shed this magnetic attraction to the cold north pole of jargon? What if they could make a warm connection with their audiences and enjoy the pleasure of a natural, conversational tone, as they had with me? And what if they could do it effortlessly, without someone standing next to them, drawing it out of them?

As soon as I asked the question, I remembered what had made our conversations so lively: I had gone back to my roots in improvisation. So now the question was, *Could scientists become more personal, more available to their audience if they studied improvisation?*

That seemed like a wild idea, but I was convinced from my own experience that improvisation was a powerful tool and it might at least be worth trying.

So I decided to experiment.

CHAPTER 2

Theater Games with Engineers

The science writer K. C. Cole invited me to have a public conversation with her about science communication at the University of Southern California, where she taught. I told her I wanted to try something, although I didn't know if it would work. I asked if she could invite about twenty engineering students to join me after our talk for an afternoon of improv.

I said that each of them should be prepared to speak to the rest of the group about some aspect of their work for two minutes. Then we would improvise for a while, and after that, they would give their talks again. I was hoping for the best, but I had no idea what would happen.

The students came into a large, empty room looking both game and a little apprehensive. When they first gave their talks, they ranged from fairly lively to rigidly stilted, which was to be expected, since none of them had had any training in communication. Some buried themselves in PowerPoint and most of them drifted effortlessly into jargon. Jargon is all right as long as the people you're talking to know exactly what you mean, but even though they were all engineering students, they worked in different areas and didn't share the same technical vocabulary. Even among fellow engineering students, some technical words were verbal roadblocks.

After their talks, I put them through three hours of basic improvising games and exercises. They found the games fun and not at all scary. Although comedy was not the objective, there was plenty of laughter and the joy of spontaneity. It looked so enjoyable that the young man who had been assigned the job of shooting a video of the session kept abandoning his camera to join the games.

There are hundreds of those games in Spolin's book, so for a three-hour session I had to choose just a few.

At first, I asked them to walk around the room, feeling the space they were walking through as though it were a real substance. "Feel it against your face," I would call out. "Feel it against your ankles." They weren't supposed to be feeling the *air* as they moved through it. Instead, this was an act of imagination in which they allowed themselves to experience something with their bodies that wasn't actually there. It was as though space itself could have substance if they just imagined it and let themselves be affected by it.

Once they had that experience, they could swoop the nonexistent space into shapes and structures—sculptures made of nothing, but that, in their imagination, they could actually see.

COMMUNICATION AS A GROUP EXPERIENCE

This group sculpture was the first step toward sharing a moment with another person, the beginning of the ability to relate.

I asked them to gather into small groups and create something *together* out of space. They had no idea what shape it would take, and because it was imaginary and created by a group, there was only one way you could see it: by observing closely how the *other* players were manipulating the space. Little by little, they saw an object emerging that they had brought into existence as a group, and it made them chuckle. No jokes, but still, laughter.

An exercise like this is the first step in bringing about an awareness that the other person is a crucial partner in relating. If another player creates a bump in the sculpture, you don't ignore it; you acknowledge the bump and build on it.

Communication doesn't take place because you tell somebody something. It takes place when you observe them closely and track their ability to follow you. Like making a sculpture out of space, communication is a group experience.

In another game we played that day, someone would start a physical activity, and the others would have to guess what the activity was. They weren't allowed to simply say what it was, but instead had to demonstrate they knew by joining in.

Someone got up and started playing an imaginary trombone. As soon as another player realized what was happening, she joined in, playing a cello. Then another player started pounding a set of imaginary timpani, and soon a whole orchestra was playing.

This led to games that involved talking and listening to one another. Step by step, I was leading the players to a heightened ability to observe one another, to be comfortable in one another's presence, and to relate in a personal way.

The question was, when these engineering students gave their talks again, would they relate to their audience with anything like the same ease that they had related to one another during the games? Or would they fall back into jargon and lecture mode? The results shocked everyone in the room.

Including me.

I had expected one or two of them to show a little progress. Instead, almost everyone had improved, no matter at what level they had started.

One student who at first had been so self-conscious that she could only look over the heads of her audience could now look them in the eye. Another had started out earlier pretty confident and polished, but had been so faithfully married to his PowerPoint that he had given most of his talk facing his own slides. After improvising, he was able to put down his remote control and speak from the heart.

Afterwards, K. C. Cole admitted to me, a little sheepishly, that she had thought it was going to be a total bust. There didn't seem to be any way that three hours of playing games could lead to the change we had just seen in the way the engineering students were able to communicate about their work.

I knew from experience that after just one afternoon of improv games, the students' progress couldn't be long-lasting. Improvising transforms you, but it does that over time. Still, we had seen something unexpectedly encouraging, and a little exciting. Even guarded, cautious engineering students could be taught to open up, to reveal their own warm humanity—to connect with their audience and speak in a way they never had before. And the flushed, happy faces of the young scientists showed that they had enjoyed doing it.

As I got in my car and left USC, I didn't know where all this would lead. But I had a vague, jumpy feeling that something important had happened—maybe even something that could change things in a way that up until now I had been just dreaming about.

In those moments on *Scientific American Frontiers* when I had pressed scientists until they finally described their work in terms I could understand, it was always a thrilling experience. I wanted other people to have that thrill, and I wanted scientists to have the pleasure of seeing people *get it;* to see them be as excited as the scientists themselves were by the adventure of hunting down an understanding of nature.

For the next couple of years, whenever I found myself at a university that taught science, I'd try to spend a few informal minutes with the president of the school and earnestly bring up my modest proposal: "Do you think it would be possible to train students in the skills of communication while they're learning science? All kinds of communication: speaking to an audience, talking to Congress, writing books, articles, op-ed pieces for the general public. Do you think your university could turn out accomplished scientists who are also accomplished communicators?"

One day at lunch, I pitched my idea to the president of a school that had produced some great scientists. He didn't sound interested. "We already have a contest where we give a prize for the best presentation."

"That's good," I said, "but a prize rewards the people who are already good at communicating. Wouldn't it be helpful to give some actual training to students who aren't naturally good at it?" I could feel I wasn't connecting. He seemed more interested in his salad than in rearranging his curriculum.

"We have too much science to teach," he said.

I wasn't sharp enough in that moment to ask a critical question: Isn't communication really essential to science? Can you really do science without communication? Will science be funded if the funders can't understand what they're supposed to be funding? Will young people choose to study science if they don't hear from scientists themselves how exciting it is?

I was still new at this and wasn't able in the wisp of a moment to raise these questions, and so we moved on to dessert.

I couldn't really blame him. He did have a lot of science to teach, and he seemed to feel that students would pick up the fine points of communication just by listening to good communicators. But just listening to good communicators doesn't work. It takes training to learn how to do it. I've been listening to good pianists all my life and I still can't play the piano.

It seemed to me that he was leaving good communication to chance—although physics, math, and chemistry aren't left to chance. But I didn't press my point. I would save my superb rhetoric for the president of the next university.

Then one night I was sitting at dinner next to Shirley Strum Kenny, the president of Stony Brook University on Long Island. By this time, I could make my case a little better, and more than that, she had come from a background in the humanities. She was interested in helping scientists communicate better and was sympathetic to the idea.

A mutual friend, Liz Robbins, is a lobbyist for causes she believes in, and science communication is one of them. She set up a meeting, and a dozen or so professors and department heads gathered in rocking chairs and swings on her back porch on Long Island. Over a glass of lemonade, they listened as I made a passionate plea about communicating science. I could see by the flicker of interest in their faces that they were actually listening. One of them was Howard Schneider, the dean of the Department of Journalism. Howie is a slim, affable tornado of energy. He left the meeting and immediately began to set up the Center for Communicating Science at Stony Brook University. Howie told me recently that the Center has become known for doing such good work that many people are proud to say they were on the porch that day. "Sometimes," Howie said, "when they weren't actually there."

But we were just getting started and we needed funding. Liz Robbins set up a meeting between me and Congressman Steve Israel, in the hope that we could get a government grant to get the ball rolling.

I sat with Steve in Liz's wood-paneled den in Manhattan and started to pour my heart out to him about how urgent it was to help scientists to be clear. He stopped me in midsentence.

"You have no idea how bad it is," Steve said. "I was at a meeting where members were all lined up on one side of the table and scientists were on the other side, explaining what they needed money for. Nobody could understand them. People were passing notes to one another. *Notes.* 'Do you know what this guy is saying?' 'No, do you?'"

He didn't need convincing. He went to work on it, and a few months later we got a much-needed grant.

We were on our way.

CHAPTER 3

The Heart and Head
of Communication

EMPATHY AND THEORY OF MIND

In 1786, Thomas Jefferson wrote to a woman he was recklessly infatuated with: Maria Cosway. He was reckless because, although she was talented, intelligent, and beautiful, he was single and she wasn't. His letter has come to be known as Jefferson's dialogue between Head and Heart. His head keeps explaining to his heart why this romance is impossible, while his heart has a mind of its own.

For me, Jefferson's dialogue between his head and his heart captures exactly what we have to be aware of when we try to communicate with someone. We have to know what they're thinking and feeling.

In a way, the dialogue is a charming argument between two horses, tethered to each other in Jefferson's mind, both pulling in different directions, and each resenting the other's influence.

After Jefferson says goodbye to Maria and her husband as the couple heads back to England, the Head and Heart squabble about the pain brought on by having struck up a friendship with this charming husband and wife in the first place.

> *Heart:* I am indeed the most wretched of all earthly be-ings. . . .
> *Head:* . . . This is one of the scrapes into which you are ever leading us.
> *Heart:* . . . Sir, this acquaintance was not the consequence of my doings. It was one of your projects which threw us in the way of it.

Like Didi and Gogo in Samuel Beckett's *Waiting for Godot,* these two aspects of one person's mind bat their arguments back and forth as if working a shuttlecock, with neither wanting to lose the point.

This interplay of the emotional and the rational, I believe, is happening in the other person's mind when we try to explain something difficult, like science, or something that's just diffi-cult to absorb, like bad medical news.

For someone to understand us, their mind and heart have to work together and resolve that inner conflict, and resolving that conflict is just what Jefferson's heart proposes to his head at the end of the dialogue. Heart is determined to continue in his affec-tion for the couple, and if Head wants to help out in keeping the couple entertained, he's welcome to try. And in return, Heart will help keep relations cordial with the scientists that Head val-ues so much." So Jefferson ends the dialogue with Heart's re-questing a collaboration, saying in effect, Help me, Head, in carrying out my softer aims and I'll help you soften the edges of your science.

It's not only a peace treaty between emotion and reason, it's a good metaphor for our strategy, which enlists emotion to help communicate difficult subjects in a more human way. Attention to Head and Heart is at the very core of communication.

Being truly connected to the other person happens when we see them in a way that's both emotional and rational, especially if we include listening with our eyes: looking for clues in the face, in gestures—in all the nonverbal signs of a state of mind. It's complete and total listening.

First, it's understanding what another person is feeling—what's usually called empathy—and second, an awareness of what another person is thinking—what scientists call Theory of Mind.

The two faculties are different, although sometimes they overlap, and some writers use the terms almost interchangeably. I think of them as dealing mainly with two different states of mind: one mostly emotional (empathy) and the other mostly rational (Theory of Mind).

First the emotional:

EMPATHY

For years, I was a little leery of the notion of empathy. If I thought about it at all, it seemed to be a free-floating kind of sympathy. Or even a semi-fraudulent posture where you announce that you feel people's pain but don't necessarily feel anything. There seemed to be something slightly New Agey about it that was more rooted in a warm wish for community than in reality. And yet, we were getting these rich experiences in improv classes that seemed to stem from developing a greater awareness of the other person's emotional state. The more I learned about what the research scientists were doing, the more I came to see empathy as a necessary part of communication.

The other person might not give voice to their feelings—their excitement, their confusion, their disapproval—but it's important to know what those feelings are. And I think this is true whether you're a scientist talking to a lay audience, a doctor talking to a patient, a supervisor talking to an employee, or a parent to a child.

But where, exactly, does this ability to read emotions (and minds, as well) come from? There's a debate in science about that.

In the 1980s and '90s, scientists in Italy discovered that certain neurons were firing in the brain of a monkey when he saw another monkey grasp something. These were the same neurons that would fire if the first monkey were grasping something himself. But he wasn't grasping anything; he was just looking at another monkey who *was* doing it. His neurons were acting as though he were performing the action, too. The neurons became known as mirror neurons and began to get a lot of attention. Marco Iacoboni, a friend of the Italian researchers, did a great deal of research on mirror neurons and became convinced of their importance in human empathy. Marco was at the University of California at Los Angeles when I interviewed him in 2009 for a miniseries called *The Human Spark*.

As we sat near an outdoor café in the warm sun, Marco explained how he felt these special neurons make reading the minds of others possible. His research suggests that when I see you grasp something, like a cup, not only do the same neurons fire in my brain as they do in yours, I even know what you intend to do with the cup. Just by watching, he suggests, my neurons mirror yours—and I get insight into what you're going to do next.

The idea is that the brain can see the action in context and quickly pick out the action that's most likely to come next. As Marco said that day, "It's all about simulating the intentions of others. We're imitating internally what other people are *planning* to do. This is important because we need to predict what other people are planning to do."

"Why?" I asked. "Why is that so important?"

"Let's take the two of us. We're having a conversation about the brain. And I can see all your body language. I have a lot of information that allows me to predict what you're going to do next. If I'm unable to read you, this conversation would be much more unsettling to me, because I'll never know if you're about to slap me in the face."

According to Iacoboni, the work of these mirror neurons doesn't stop with reading the actions and intentions of others, but also extends to reading their emotions. He's written that mirror neurons send signals to the emotional centers of the brain, and that's why we react to the facial expressions of others in an appropriate way—"the happiness associated with a smile, the sadness associated with a frown. Only after we feel the emotions internally are we able to explicitly recognize them."

He says, "When we see someone else suffering or experiencing pain, mirror neurons help us to read her or his facial expression and make us viscerally feel the suffering or the pain of the other person." According to Iacoboni, those moments "are the foundation of empathy."

This all sounded exciting when Marco and I were having an animated conversation at UCLA, but since then, some clouds have drifted over the subject of mirror neurons. There are neuroscientists now who are not convinced that mirror neurons enable us to predict what a person's intention is, or even if they exist in humans. According to them, seeing an action and then figuring out what's coming next may be the work of other networks in the brain.

This is clearly not a debate I'm qualified to weigh in on. But

I'm not unhappy to see it happening, because it shows how science works. Somebody observes something that hasn't been seen before and draws a conclusion about what's actually going on. Somebody else isn't convinced and they argue about it. But instead of arguing endlessly, they continue to do research, hoping to find out what's really happening. While they might sound antagonistic, they're actually involved in a kind of messy collaboration to understand something fundamental about nature. They may very well find that neither side is right and instead discover another mechanism that no one had yet thought of.

Whatever it is, it would be good to know what's causing empathy. As one scientist said to me, "Certainly, we must have *some* system in our brain for creating empathy. Because we do it."

Empathy gives us a sense of what's going on in what Jefferson called Heart, but we also need to be aware of what Head is up to—what the other person is *thinking*.

THEORY OF MIND

When I first heard the term *Theory of Mind,* the idea was a little difficult to grasp. I had always thought, naïvely I guess, that we all realize everyone else has their own private thoughts, which are different from ours. But in talking to scientists on an episode of *Scientific American Frontiers,* I learned that young children don't think that way. I watched experiment after experiment with children below the age of four and a half or five in which they were sure that what *they* knew was also known by everyone else. For instance, if a very young child watches a cartoon in which a woman enters a room, puts a cookie on a table, and leaves, the child clearly understands that the woman knows she left the cookie on the table. But if, while the woman is gone, a man enters the room, picks up the cookie, hides it in a cupboard, and leaves, the child suddenly changes her understanding of what the woman knows. Now the child thinks the woman knows what *the child* knows: that the cookie has been moved.

If the woman comes back into the room and you ask the child, "Where does the lady think the cookie is?" the child will point to the cupboard. The child saw the cookie go into the cupboard, and it doesn't occur to her that the woman doesn't know what the child herself knows. The child has no theory of what's going on in the other person's mind, other than that it must be the same as what's going on in hers.

This is a natural stage in a child's development. In fact, it's not until about the age of four or five that it even occurs to children that deception is possible. There's no point in lying if everybody knows what you're thinking!

But once Theory of Mind develops in a child, it also becomes clear that others might be lying to *you,* and it's kind of important to know what's going on inside these other people's heads. So once we have Theory of Mind, it's a tool we rely on all our lives. We wouldn't fork over money on a used-car lot without trying to figure out what the salesman is thinking. *Is this person hiding something about the mileage gauge? What's his agenda? Is my welfare at stake here?*

In a situation like this, our understanding of what's going on in the other person's mind is the result of several kinds of listening. We put together all the clues we can—facial expressions, tone of voice, body language, and any telling words they let drop.

This is exactly what we do in the improv classes: observe the other person, track their body language and tone of voice, intuit their thoughts and feelings.

And suddenly, we're reading minds.

CHAPTER 4

The Mirror Exercise

Two people are standing face-to-face. They look intently into each other's eyes.

Very slowly, one of them starts to move, and immediately the other moves as if she were his mirror. Soon they're in a kind of slow-motion dance, in which the motions of each one are an exact mirror image of the motions of the other, with no lag time: a perfect mirror.

These two young scientists have agreed, along with a dozen others, to take part in another informal experiment, this time at Stony Brook University. I want to see what effect three hours of playing improv games once a week for six weeks will have on them. Will they become better at relating to others and therefore better communicators? If so, will it stick?

As they do this mirroring exercise, they'll each learn to observe the other person so closely that eventually, almost without thinking about it, they'll be able to predict what that person will do next.

Silently, the leader does simple things, like stretching out his arms or scratching his face, and the follower has to observe his slightest movement so well that she can anticipate what comes next and do it at the exact same time he does it. The motions of both of them should be so in sync that a stranger walking into the room wouldn't be able to tell who was leading and who was the mirror.

At first, they're having trouble. The person playing the mirror lags behind. She's not so much a mirror as a delayed video recording of his movements. But that's because he's not giving her a chance to keep up with him. He's moving too fast. I coach from the side and explain that it's *his* responsibility to help the mirror, his partner, keep up with him. This is the students' first glimmer of the basic idea: *The person who's communicating something is responsible for how well the other person follows him.*

If I'm trying to explain something and you don't follow me, it's not simply your job to catch up. It's my job to slow down. This is at the heart of communicating: If I tell you something without making sure you got it, did I really communicate anything? Was I talking to *you*, or was I just making noises? In the mirror exercise, is the leader enabling the follower to follow, or is he just waving his arms?

I ask the leader to slow down and give his partner a chance to follow, to connect with him, and before long they're actually in sync.

Then the exercise gets a little tougher. I tell them that the follower, the mirror, is now the leader.

They struggle a little, but after a few minutes they're pretty good at this switch in roles. A stranger walking into the room would have a hard time knowing who is the follower and who is the leader.

But then I throw them another curve. I give them an instruction that has been mystifying people new to improvising ever since Viola Spolin invented her "Theater Games" decades ago. I tell the players that now neither one of them is leading; they have to find the motion together. They're both leaders and they're both followers. This doesn't just seem hard to do, it sounds impossible. Somehow, they have to achieve a kind of instantaneous synchrony.

They work at it for a while, and before long the young scientists are surprised to find they can actually sync up, even if no one is leading. There is delighted laughter. An expression of surprise, even a little shock.

They're beginning to read each other's bodies, learning to pick up clues that will lead eventually to reading each other's feelings and thoughts—and in this way it will be as though they're what's called "reading each other's minds."

We move on to something even more difficult.

VERBAL SYNC

Now the couple sits in chairs, facing each other. Instead of mirroring their movements, they're going to be mirroring one another's speech. I ask one of them to tell the other about something ordinary—what she did to start the day, or the plot of a movie she saw. She has to improvise it on the spot. With no preparation of any kind, their task is to be so in sync that *they both say the exact same thing at the exact same moment.*

This isn't easy. They begin with huge lags between leader and follower. Coaching on the side, I tell them, "Speak at the *same time.* You can't be an echo. You have to be her mirror. A mirror does what you do at the exact same moment." I urge them to slow down and give the other person a chance to keep up. You can see their frustration, but they stay focused on each other's eyes, lips, bodies. Listening intently, they try to pick up signals of what the other might say next.

> *He:* This is really hard! [*laughter from the class*]
> *She:* No . . .
> *Both:* . . . kidding.
> *She:* It makes the whole . . .
> *Both:* . . . science . . . thing look really easy. [*more laughter*]

And they're in sync.

But how can a simple game like this have anything to do with good communication when these people are out in the world? It's a good question. Most of the scientists we work with are adventurous, but at this point, some express skepticism. "Okay, this is fun, but what is it actually doing for us?"

It wasn't that easy to explain at first. I assured the doubters that by trial and error we had found that it worked in real life. "It helps you connect to the other person, to truly relate to them," I said. If the students would just stick with it, learning to soak up the clues coming to them from the other person's body language and tone of voice, learning to read the other person, they'd see themselves becoming better communicators. I'd seen it happen. But anecdotal evidence like that is not completely satisfying to a skeptical scientist. I wondered if there was any hard evidence to support these improv games. I started reading research papers in Stony Brook's online library catalog, and discovered that there *has* been scientific research on behavior similar to the kind of exercises we were doing.

MARCHING AND TAPPING

At Stanford University, Scott Wiltermuth and Chip Heath wondered why modern armies still train by marching in step and shouting cadences. Marching in step toward the enemy has long been out of style, largely because it tends to be suicidal. So why does synchronous marching persist in the training of soldiers? Is there some advantage to it?

The answer, they found, is yes. It can strengthen cohesion and promote cooperation within a group, and here's how they figured that out.

In one study, they had small groups of people walk around the campus. Some groups were walking in step, while others, acting as a control, were walking normally. After the walk, the participants took part in a game that is known to test a group for trust and cooperation—and those who had marched in step scored higher on both counts.

I had to read that result more than once. *The simple act of walking in step produced greater cooperation? And more trust?* It seemed hard to believe. But the tests of the groups' cohesion were standardized. Their reliability had been verified many times over.

In other studies, simply *tapping* in sync, like tapping on a table, produced the same results. After they had spent some time tapping in sync, the subjects paid more attention to the good of the group, and they made fewer selfish choices.

Piercarlo Valdesolo and David DeSteno found that synchronous tapping gave participants an increased sense of similarity with the person they were syncing up with. The researchers reported that with that feeling of similarity, the subjects were more likely to have compassion for their tapping partners and to behave more altruistically toward them.

I want to be cautious and not regard these, or any of the studies I write about in this book, as the last word in understanding human interactions. Most research, I think, can only suggest, or point toward, an insight of some kind. One thing you can say for sure about most studies is that they leave you with the suggestion to do more studies. But the research I've read does seem to throw light on what I've experienced many times in improv sessions.

For instance, marching and tapping in sync is very similar to what we do in the mirroring exercises. And the scientists who did these studies found some of the same bonding of the participants that I had observed in our improv classes.

As a young actor, when I worked with Paul Sills for six months on Theater Games, I felt a sense of camaraderie I hadn't known before. Now I was seeing controlled studies that suggested this feeling had come from syncing up with the other players.

LEADERLESS SYNC

Right around the time our grad students were trying to sync up without a leader on a stage at Stony Brook, a team of scientists in Israel was doing experiments to see if such a thing was actually possible.

Uri Alon, along with Lior Noy and Erez Dekel, at the Weizmann Institute of Science, had worked out a method to test how well people could mirror one another.

Uri is a scientist who, in addition to his research, also runs workshops for younger scientists. He works with people who are just making a big transition in their lives, where they go from postdoc to running their own labs and becoming group leaders. "It's a time," he says, "when they're most panicked. They need to do things they've never been trained to do, and they make many basic mistakes in communication."

They're sometimes uncomfortable being the boss; they have trouble figuring out how to guide people and often don't know how to listen.

Uri has a background in theater and has taught improvisation. He was sure that Theater Games could help the scientists open up and connect to the teams they'd be building. But he hit a snag when he asked fellow scientists to do leaderless mirroring.

The exercise has been practiced by actors since Viola Spolin created her Theater Games many years ago, but the scientists hadn't heard of it and didn't believe that leaderless mirroring was possible. They thought that *somebody* had to be leading at each moment. And even when there appeared to be no leader, they were sure that, imperceptibly, the players were probably taking turns.

Curious, Uri set up a study to see if something, in fact, does happen in mirroring that can bring two people so close mentally that they can anticipate each other's movements spontaneously, apparently without either person's leading—and within milliseconds.

He set up his study with the subjects facing one another, but instead of moving their bodies, which would have been difficult to measure, he asked them each to move a handle along a track. If one of them pushed his handle forward or backward, the other person would try to mimic the movement with *her* handle in the same direction and in the same moment. The movements would be recorded on a graph, and in this way Uri could measure any lags in mirroring in milliseconds.

They found distinct patterns in the way the participants moved the handles, and among experienced improvisers these patterns showed a striking result. When they engaged in a purely joint improvisation with no leader, their mirroring was *more synchronized and more rapid* than when one of them led and the other followed. They were actually *more in sync* when there was no leader.

So, I was beginning to glimpse something at the heart of improv—why it creates such a strong group experience. Everyone who has improvised for a while senses it: Synchrony brings us together.

CHAPTER 5

Observation Games

Synchrony can't occur without acute observation—which is why improv training often begins with games and exercises that enable people to see and respond to the slightest shift in another person's behavior.

For instance, once people are used to the idea of working with the imaginary substance of space—walking around in it and building sculptures out of pure air—they'll stand in a circle and one person will shape an object in her hands and pass it to the person next to her. He has to observe how she handles the object to know what it is. She'll use an egg beater differently from a golf club, but his only clue is observing her motions as she uses it. When he takes it from her, he has to handle it in a way that shows he knows exactly what it is, including its size and weight. Because it's invisible, it exists only by common agreement. Suddenly, everyone is seeing things that weren't there before. It's sort of like what democracy could be if we actually paid attention to one another.

You see this when they toss imaginary balls back and forth. When someone receives a ball with the same force with which it was thrown, and with the same weight and size it had when it left the other person's hand, you can actually see a ball going back and forth.

I've seen astonishment on people's faces as they watch a volleyball game with five people on each side of a net that isn't there. They're batting a nonexistent ball back and forth and yet they can keep score. They're absolutely clear about who won or lost a point, because they can see precisely where the imaginary ball is at every moment.

It's just as surprising to see two opposing teams pick up an imaginary rope and have a tug-of-war with it.

When one team pulls on the rope, the other team has to know exactly how much force is applied, and how much resistance their own team is coming up with. Above all, they have to make sure that the rope between them stays the same length. It can't suddenly be made of rubber and get longer or shorter. They do this by watching for the slightest movement in the other team's bodies, but also in the bodies of their own team; how much are they countering the efforts of the other team? Each person has to sense where they are in a complicated mix of forces. Somehow, they have to add their own force to it, but they can't control the outcome all by themselves. It's a large group responding to subtle changes, and it's leaderless.

First, the team on the left is pulling so hard that the one on the right starts being moved across the floor. Then the players on the right dig in their heels. After a moment, the tide turns. The team on the right is able to pull the opposing team toward them. They're all straining against the taut rope until, finally, one team can't resist anymore and they fall over one another, collapsing on the floor.

And there was no rope. It only became real because they were observing one another and accepting the dynamics of how the group handled it. As soon as they let go and collapsed on the floor in laughter, the rope disappeared.

GIBBERISH

Observation games prepare the players for more complicated games, where *words* are used to communicate something, but they're not always words found in the English language. If you walked into the room during one of our workshops, you'd see some weird things.

Someone might be selling a product to the class, but the class has to work hard to guess what the product is, because the sales pitch is entirely in gibberish—nonsense sounds that sound like a language but have no meaning.

It takes a while to figure out your own version of gibberish. At first, some people are intimidated by trying to speak gibberish, because it feels strange to make garbled utterances when you're really trying to express a thought; you want the actual words of your native language to come out, not these strange sounds. But here you are selling us a product and you have to *show* us why we should buy it—not through words, but through body language, through the way you demonstrate it, and how emotional you are about it.

Before you know it, you're communicating with your whole body. And in that moment, you have the pleasurable experience of making contact with people, not by spraying words at them, but by using your whole expressive self. After a while, we actually see what you're trying to sell us. And we buy it.

WHAT'S THE RELATIONSHIP?

In another game, someone sits in a chair, waiting. After a moment, someone enters the room and begins explaining something to the person in the chair. Without any clues—just by observing her manner in how she relates to him—he has to figure out what the emotional relationship is that he shares with her. Is he her beloved little brother? Her strict father? Her overbearing boss? If she's a scientist, she might tell him about her work in the lab, or she might tell him the plot of a movie she's seen. It doesn't matter what she talks about. Her job is to convey their relationship, not through what she says, but only through her manner, through how she relates to him.

This is a variation of Viola Spolin's intriguing game called Who Am I? In our version, a researcher might be telling someone about her work in science but, as in Spolin's game, through her tone of voice, her body language, and the way she phrases her statements, she'll be trying to let him know who he is in relation to her. She would talk to her boss with different words and a different tone than she would use with a child. If she's trying to convey to the other person that he's playing the part of her nine-year-old brother, she'll have a hard time doing it if she rattles off a couple of minutes of jargon. But it's against the rules of the game to use hints. For example, she can't mention "our mom" or ask, "How was school today?"

As she launches into the conversation, she's faced with the realization that her usual way of talking about her research is probably not going to convey to him who he is. One thing she'll learn from this is that there are many different ways to express the same thought, depending on whom you're talking with. This can help if one day a member of Congress doesn't get it the *first* way a scientist explains her work and she has to come up with another way to say it.

More importantly, she'll learn that a lot of communication takes place in ways that don't use words. The scientist is talking to the other player about her day in the lab, but what she's really trying to communicate to him is their relationship— that he's her little brother. He's completely in the dark until he picks up the right clues, not from any hints she's dropping, but from the way she relates to him. He's learning to read her behavior. And she's learning to express herself through her behavior.

It sounds like you might need actors to do this well, but I've been surprised to see how vivid people can be when they simply focus on the person in front of them.

I once saw Dr. Martha Furie, a professor of pathology, come through the door and brusquely confront another woman. As Furie began talking about her work on Lyme disease, everyone in the room could tell, judging by the sight of her angry face and the cold tone of her voice, that something big was at stake. It certainly was: The imaginary relationship I had given Martha was that the other player was her estranged sister who was having an affair with Martha's husband. It was a pretty fraught relationship, but the training she had had, the games she had already played, enabled her now to abandon herself to this bizarre imaginary situation. She was even able to do it knowing that a reporter from the *New York Times* happened to be watching the session that day. "That was crazy," Dr. Furie said later to the reporter. "I'm actually not a person who puts myself out there. I can't believe I did that."

This kind of thing happens often in improv training. People open up in ways they never expected.

Even trained actors can open up in unexpected ways.

I was invited to run a master class for actors at Ten Chimneys, a museum in Wisconsin that was once the home of the celebrated acting couple Alfred Lunt and Lynn Fontanne. I had already been spending a lot of time working with scientists, and it had been years since I'd taught improvising to actors, so I was eager to get together with them.

They were maybe a little less eager than I was. They all had twenty or thirty years' experience on the stage, but when they heard we were going to improvise for a whole week, they showed a slight air of hesitation, in some cases bordering on panic. One especially gifted actress admitted later that she was so determined not to improvise that she was thinking of faking a heart attack.

I wouldn't have known this unless she told me, because she made such an amazing leap in her work during one improvisation. The imaginary circumstances were as fraught as they had been with Dr. Furie.

The actors were reading a scene from the Friedrich Dürrenmatt play *The Visit*. The actress played an older woman who comes back to her village to confront the man who had impregnated and abandoned her when she was a young woman. The scene calls for her to be full of unexpressed hatred. In the reading, both actors were committed and believable, but the idea was to see if we could go deeper using improv. What if we went back to the primal moment in their lives that prompted this scene, when they were in that place in the woods where he had abandoned her thirty or forty years earlier? This scene is not in the play, but I asked them to improvise the event when she realizes he intends to shrug her off and she begs him to marry her.

At first, they stood looking at each other, just making up dialogue, as most everyone does to avoid being in the moment. When this happens, the words sound invented, unspontaneous—because when you're making up dialogue, you're thinking of what you'll say next; you're not focusing on the other person and reacting to them. I needed to get them to focus on something outside their own heads, so, coaching from the side, I asked them to use the imaginary environment, to make contact with the *place*.

Within a few seconds, their focus was on where they were and on each other, not on themselves. Suddenly, they noticed it had begun to rain. When she asked him plaintively, "Don't you want to marry me?" his answer was to wipe a raindrop off her nose. He laughed as if the raindrop were a funny little distraction, which drove her into despair. Real anguish came out of her.

When the improv was over, they again read the actual scene written in the play. The first reading had been good; this time it was startling. Her tone now was venomous, her hatred far deeper than during the first reading. She wasn't working from an intellectual understanding of the circumstances; she had actually just lived through a traumatic event from the past and it had changed her. The actress herself was surprised at how far she had been propelled by the improvisation. "I would've arrived there eventually," she said. "But not like this, in ten minutes!"

At Stony Brook, Dr. Furie had the same experience, and no less so because she wasn't an actress. Acting wasn't involved—at least not what people conventionally think of as acting. There's no pretending in improvising. No *deciding* to behave differently. Instead, with our attention on the other person, and with a heightened ability to respond, we're tuned in to the present moment; we're in intimate contact with each other. It's as though the raw, vulnerable tissue of our most private self is in contact with the same vulnerable self of the other player.

We can sense what they're feeling, and we have a greater awareness of what we ourselves are feeling.

This is what is usually called empathy.

I've come to see this connection with the other person as the bedrock of communicating. It's surprising how effective this ability to tune in to others can be.

43

CHAPTER 6

Making It Clear
and Vivid

Bob Conn, the president of the Kavli Foundation, sat me down at breakfast one day and said, "Look, you can't just work with graduate students. Most of them won't be communicating with the public for years. There's a crisis right now in communicating science. You have to start working with senior scientists. They're out there now." He was right.

His foundation had been funding some of our workshops around the country, and his advice paid off when we went to the Kavli Institute for Nanoscience at Cornell and for the first time worked with senior scientists who were already communicating with the public. David Muller was one of them.

David had recently made an important breakthrough in nanoscience. And while we were there, he made a breakthrough in communication, too.

A few months earlier, David and a grad student had created the world's thinnest piece of glass: only one molecule thick.

They had achieved something extraordinary—a new way to understand the structure of glass—and their accomplishment was published in a few science journals.

In our workshop, though, Muller discovered a new way of *talking* about the glass.

As he told me later in an email, "The workshop was enormously helpful in helping us craft an effective story and understanding what did and did not resonate with folks."

That was probably because we actually gave the scientists real folks to resonate with. We had invited three nonscientists into the classroom, three intelligent people, from average walks of life. David and his fellow nanoscientists had to make clear to them what two-dimensional glass was all about. Before they could do that, they had to make them *want* to know about it.

As David reminded me later, "This was something you keyed in on, when we were discussing the abstract. I think you said something like 'Wait, that's very unusual for a scientific paper, tell me about it.'"

David and his student had made their discovery *by accident*. That had caught my ear, and I thought other nonscientists would find it interesting, too. While David and the student were working with a thin layer of graphene, an air leak had produced what they thought was "muck" but turned out to be glass. David realized he could begin his story with that very human moment of accidental discovery.

"That helped me understand where people were and which ideas made good entry points to a discussion," he said. "The accidental nature of the discovery helped people make an emotional connection, and I think that's why it was picked up more widely than I would have otherwise expected. It is something we would not have emphasized at all in a straight retelling, but once we told the backstory, people got pretty excited."

Not only did he now have the human element of an accident to talk about, he had something else. Between the discovery of the glass and our workshop, his lab had been placed in *Guinness World Records* as the creators of the world's first two-dimensional glass, giving him another popular hook on which to hang his story.

With this new insight into what would capture the attention of laypeople, the next time he spoke to an interviewer about the world's thinnest glass, the story was picked up by websites and newspapers all over the United States and the United Kingdom. He was able to talk about his discovery not just in terms that interested him as a scientist, but in a way that ordinary people would find interesting.

He was even contacted by venture capital firms who had seen the media stories and were asking if he wanted to commercialize the glass. It was too early in the research cycle to take advantage of those offers, but, for me, David Muller's story is an example of how good communication can make the difference between being noticed by a few technical journals and getting the attention of much of the rest of the world.

David was making use of an increased ability to know what was in the mind of his audience as he told them his story.

Just as in the mirror games, where the leader is responsible for how well the other person follows, he was taking responsibility for how well his listeners were following him.

The surprising thing about this is that empathy and Theory of Mind—strengths that make this possible and that most of us seem to have naturally—are not called into play as much as they could be.

Too often, reading minds correctly is just beyond our grasp, when in fact it can be one of our most powerful tools.

CHAPTER 7

Reading Minds:
Helen Riess and Matt Lerner

I have a friend who went to a doctor with pains in her foot. The doctor was kind and sympathetic, and when my friend described her symptoms, he dropped his head into his hands and said in an anguished voice, "Oh my God, it sounds like *plantar fasciitis*!" His tone was so miserable that my friend was afraid she had an incurable disease, rather than a pain in her foot that would probably go away with the right treatment. He had empathy, all right, but he was so swamped by feeling her pain himself that he left her out of the interaction. It wasn't all that helpful.

Most people agree that empathy involves putting yourself in the other person's shoes and understanding in some way how they feel. But why would empathy lead to the doctor's scaring his patient like that? I needed some help in understanding this and started searching for someone to teach me more about empathy. A TED talk online led me to exactly the right person.

HELEN RIESS AND DOCTOR-PATIENT EMPATHY

We sat at my dining room table with a cup of tea, sharing some berries and a bowl of nuts. Helen Riess had agreed to drop by and tell me about her work at Boston's Massachusetts General Hospital. Helen is a psychiatrist who is helping to revolutionize the way doctors relate to their patients. I told her that I had initially been a little leery about the whole notion of empathy. She looked surprised. She said, "I want to tell you why I got so excited about teaching empathy."

Helen was asked to take part in a study of doctors during therapy sessions with their patients. The study was the idea of one of her students who wanted to see if the doctors and patients were really in sync. To test this, both doctor and patient would be videotaped in sessions *and* they would both be hooked up to skin conductance monitors. These are the devices that measure the minute amounts of sweat your skin gives off during moments of tension, or, as Helen said, autonomic arousal. This is the same technology used in lie detectors—machines that don't work so well at detecting lies, but *are* good at showing emotional shifts like the ones Helen and her student were studying.

At first, Helen was reluctant to get hooked up in tandem with her patient, but when she did, "It turned out to be a career-defining decision."

The patient was a young college woman who had a problem with weight loss. She and Helen were hooked up to the skin conductance device while the video camera recorded their session. "Later that afternoon my student called up and said, 'You've got to see this.' So I went down and I looked at our tracings. I was blown away. This very self-confident–appearing woman turned out to have massive anxiety."

She realized that they were generally in sync but their tracings had different signatures. The rise and fall of Helen's tracings were smooth, whereas the young woman's had a series of rapid spikes. Helen could see in the patterns of the tracings that although she had sat across from her, she had clearly missed reading the emotions the young woman was going through. She really didn't know what was going on inside her patient. She showed the tracings to the young woman, who said, "I'm not surprised by this at all. I live with this every day, but no one has ever seen my pain." Moved by this, Helen studied the video of the session more closely.

Examining the spikes in the patient's emotions, she could see the woman was having "these little leaks," as she described them: leaks of emotion that didn't necessarily show on her face. "She was very good at concealing them." So, to see if there had been any physical signs of these emotions that she might have been able to pick up on, Helen checked the video for involuntary gestures. "I'd look to see what she was doing when she had these spikes in the tracings. And I saw these little tics, like throwing her hair back or doing this sudden chortle. I felt like an emotion detective."

Now, she was able to see things that had escaped her before. "I didn't realize how much was going on internally until I saw the tracings. There we were in therapy," Helen said, "and I had thought I understood her. She seemed very composed, but inside she was a wreck."

As I listened to Helen's story, I was surprised that no one had tracked this kind of emotional gap between doctor and patient before. After all, I thought, the technology had been around for some time. Helen explained how one additional element had made all the difference. "Back in the sixties," she said, "somebody had the idea of measuring emotions, but they were just measuring the patient." Measuring the *doctor's* emotions and tracking them alongside the patient's was the critical new idea. When she tried it out, it made a huge impact on Helen.

Now that she was more aware of her patient's feelings, Helen could work with her in a way she hadn't previously been able to. "As I paid attention to the signs and responded to them, our work went to a much deeper level," she said. Finally, Helen was able to deal with her patient's weight loss problem. "My patient unburdened herself emotionally—and she started to exercise for the first time in her life. And this woman who had only gained weight and never lost weight before went on to lose almost fifty pounds." As much of a breakthrough as this was for the patient, it was also a breakthrough for Riess: "I realized that with this careful attention, I had learned to be more empathic," she told me.

It launched Helen on an effort to teach other doctors how to be more empathic with their patients, as well.

SWAMPED WITH EMOTION: "AFFECTIVE QUICKSAND"

I wondered how Helen taught doctors to avoid some of the pitfalls in empathy. For instance, the doctor who had become so caught up in my friend's plantar fasciitis was clearly an empathic person, but had managed to scare her. He was feeling what she felt, but he seemed momentarily swamped by it. What did Helen teach doctors that prevented that?

"First, I introduce them to the idea that we're wired to read other people's minds. That we can get under a patient's skin." She took a beat. "And that we also have to get back out." Doctors, she said, can be trained to regulate their own emotional response and avoid what she calls "affective quicksand."

This was the quicksand my friend's doctor sank into when he

was so caught up in imagining the pain in her foot (pain he himself had once experienced) that he reacted as if it were happening to *him* right then. He was not regulating his own emotions and as a result was swamped with empathy.

Helen's experience after being hooked up to her patient points to the idea that empathy (without the quicksand) is a step toward a good therapeutic connection. And it seems to me that it's the kind of connection we need to establish to speak with clarity to an audience. In fact, it seems central to good communication in general.

MATT LERNER: COGNITIVE AND AFFECTIVE EMPATHY AND THE AUTISTIC SPECTRUM

There is a group of people notoriously poor at reading the minds of others: children and adults on the autistic spectrum. As I looked through the literature, I came across a researcher who had been working with kids on the spectrum for several years and getting remarkable results, and he was doing it by having them play improvising games. I had to meet him.

I visited Matt Lerner in his lab at Stony Brook University on a warm spring day that was a relief after a harsh winter, but as soon as we sat down in his office there were no pleasantries about the weather. We plunged immediately into excited talk about communication. I told him how I had come to think of empathy and Theory of Mind as fundamental to communication. "Right," he said. "Cognitive and affective empathy . . . which, by the way, is what we're doing right now. Nodding and smiling as we talk."

He told me he did a lot of work with kids on the spectrum who, while they're fine at empathy, do sometimes have a problem with more complex and subtle aspects of Theory of Mind.

"They're an example of how this fundamental task gets disrupted," he said. "But the principles really apply to everybody."

I asked how he got started working with autistic kids, and he told me about a single day that had launched him on a life of discovery when he was only twelve years old.

Matt and his younger sister visited her friend's house. Matt noticed that in the other room, a two-year-old was sitting on the floor playing by himself. The boy's mother could see that Matt was interested and explained that her son, whose name was Ben, had a problem that had something to do with autism. Curious, Matt asked if he could go in to play with the boy. She said he could, but she assured him he would be very bored. And in fact, when Matt sat down next to Ben, the boy never looked up or acknowledged his presence.

Okay, Matt thought, *he can't really get into my world, so I wonder what's going on in his world?* Ben was playing with a toy car, running it back and forth. Matt decided to mirror what the boy was doing.

"I started doing the same thing," Matt said, "running a car back and forth." He could see that Ben noticed, but still, the boy wasn't paying much attention to Matt—until Matt deliberately did a slight variation on a motion Ben had made to see if the boy would notice it. And he did. This was the first, slight moment of connection between them.

Matt was with Ben for the next two hours, following him around the house, mirroring his activity, sometimes varying it slightly to get the boy's attention. At the end of the day, as Matt left, Ben looked up at him and said, "Matthew."

At least, that's how Ben's mother remembers it. Matt doesn't remember if the parting moment was actually that dramatic, but something surely clicked between them that day.

Matt continued to work with Ben through Matt's high school

and college years, trying to find ways to help him get along in the world. In his early years, Ben's social skills were impacted, including some very basic ones. For instance, he had trouble with beginning conversations when people said "Hello" to him. He knew a book by heart that began with the word "Hello," though, and if you said hello to him, he would answer, "Hello, Mickey. It's me, Donald. It's nice to see you again." And if you didn't stop him, he would recite the whole book.

By the time Matt was in college, Ben was enrolled in a class where he was being taught how to respond to situations he might encounter in public. But he was being taught specific, rote responses, and as Matt realized, the world doesn't always work in such a predictable way that you can have a rigid set of responses to it. So Matt started devising variations on Ben's practice sessions. He would deliberately introduce unlikely scenarios, like ordering ice cream on Mars, to throw Ben off and prepare him for the unexpectedness of real life.

After he graduated college, Matt got a summer camp job, teaching social skills to teens on the spectrum. But again, he was expected to offer them training that he felt was not going to prepare them for uncertainty. He was told to follow a manual that included things like "When you meet somebody, look them in the eye. But not for too long, so you don't scare them." The kids pretty much hated this.

"I have to tell you," Matt said, "they weren't too thrilled. Whether you have autism or not, what teenager wants to sit there and be told this is how you're supposed to behave? They were throwing things at me. They were really upset."

Matt was rooming with two actor friends at the time. They stayed up late one night and, over a glass of wine, he told them he was stuck. Matt recalls that before long, "the subject turned to improvisation, and the goals of improvisation—making eye contact, being able to quickly respond to another person's emotions, being able to resonate with them." This was what Matt had been looking for.

Recalling their own training as actors, his roommates started suggesting Theater Games that might open up the kids: mirroring, tossing imaginary balls, reading one another's body language.

"We kind of hypomanically stayed up all night, writing down different improv games I could use," Matt told me.

The next day, he began introducing the teenagers to games and exercises like mirroring and, eventually, acting out scenes from their daily lives. He saw an immediate response from them. "They were interested," he said. "They weren't throwing things."

They were no longer following a set script.

They had to follow the rules of the games, but the teens were now in control of the life experiences the games were preparing them for. They could practice spontaneity on imaginary real-world encounters that they found interesting—which weren't always totally real-world. As Matt recalls, there were "countless lightsaber fights" in imitation of the duels in *Star Wars* movies. But they were relating. By the end of the summer, they had even created and performed in their own play.

After that, Matt returned his focus to the topics he had studied at school. Music and philosophy.

But the following spring, Ben's mother called to tell Matt that things weren't working out for Ben with any of the other social-training–focused summer camps in the area. Desperately wanting help for her son, she said, "You did so well with that crazy idea last summer. Why don't *you* start a camp?" Although Matt had never done anything like that before, he managed to get a grant and hired his actor friends to teach improv. He also talked extensively with Ben to make sure that the camp would involve things he thought would be fun, rather than things he would be made to do.

Without conducting a controlled study, Matt couldn't make any claims of efficacy yet, but the kids attending his camp began showing a high level of involvement. They were writing their own plays and shooting their own movies, and their social skills were improving. At the end of the summer, the parents of all the children asked Matt to continue the program year-round. They were sure this was working.

Matt realized that the parents weren't paying him just to feel better about their kids; they were expecting to see social skills and empathy actually improve. He needed hard evidence that the techniques were working, so he did a controlled study. He and his colleagues published the first study to rigorously measure the outcomes of improv classes for teens on the spectrum.

He used multiple measures of success in the study: written and visual tests of empathy that were administered before and after improv training. And he tracked the students over time, even going into classrooms, interviewing other students to see if the kids they were studying had indeed been making friends. The program, called Spotlight, was working, and still is. Years later, Spotlight is serving over 350 children a year in the Boston area alone.

And Ben, the boy who started it all, has graduated from college and, most recently, graduate school.

It was clear that the ability to read the other person is a powerful tool. And I was about to find out that it helps not just in relating to another person, but even among whole groups of people.

CHAPTER 8

Teams

In 2010, Anita Woolley and her colleagues were trying to figure out what the most important factors were that made a team work well. It had long been established, Woolley said in her research paper, that intelligence tests could predict pretty reliably how well an individual could perform in a variety of tasks. Someone with a large vocabulary, for instance, might also be expected to be good at math. But no one had yet tested whether this was also true for *groups* of people. Was that same factor, general intelligence, at work in teams? For instance, if you did a study in which you put together a team of high-IQ people, would that team do better than other teams across a variety of tasks? Would IQ be the most important factor, or was there some other deciding factor for a team's success?

They found there *was* a deciding factor—and it wasn't IQ.

Woolley's group gathered 697 volunteers and divided them into small teams of two to five members each. They tested them at a number of tasks and found that the average intelligence of a group could not significantly predict the group's performance. What could predict it, though, were three factors: the ability of the members of the group to freely take part in discussions, members' scores on a standardized test of empathy, and, surprisingly, the presence of women in the group.

Women's presence seemed to matter at least partly because they typically have higher ratings in empathy than men do. The more women you had in the group, the higher the level of empathy the whole group had, and the better it performed. So, actually, empathy played a part in two out of the three factors that predicted success: higher scores on the empathy test and the presence of women. It's even possible, I think (although Woolley and her colleagues don't say this), that the willingness of a group's members to enter the conversation freely might *also* be related to empathy. If you get the sense that the rest of the group is aware of your feelings and won't punish you for speaking up, you might be more inclined to offer an idea. But this could be me being just a tad overenthusiastic about empathy.

The researchers' biggest surprise came when they studied groups that were not in the same room, but were communicating online: They found the same results. Whether face-to-face or online, they reported, "some teams consistently worked smarter than others." And the reasons they worked smarter were the same: They had "members who communicated a lot, participated equally, and possessed good emotion-reading skills."

On the theory that women possess a high degree of these emotional skills, at least one large company has already knowingly experimented with having a greater presence of women. Jack Ma, the founder of the Chinese web giant Alibaba, has been quoted as saying, "I feel proud that more than 34 percent of senior management are women. They really make this company's yin and yang balanced." The total workforce of Alibaba is 40 percent women. Ma says that what the women contribute is "the 'secret sauce' of the company."

This was a deliberate decision on Ma's part. But dozens of other companies have been part of a similar experiment without realizing it—and have been getting the same results. Cristian Dezső and David Gaddis Ross studied the fifteen hundred S&P firms over a fifteen-year period and found that when they had women in the top managerial positions, the firms were more successful. Interestingly, firms that had a strategy of innovation enjoyed the most success (but those that had a less innovative strategy did no worse having women at the helm). The authors suggest that the presence of women in top management positions helps specifically in situations where the focus is innovation. This, they say, is because women's social skills lead in part to better decision making overall and also because studies have shown that "gender diversity in particular facilitates creativity."

URI ALON

These researchers weren't alone in exploring how teams work best. At around the same time, on the other side of the Atlantic, a scientist in Israel was experimenting with his own techniques to build a smarter, more collaborative team.

Uri Alon and his team were the scientists who had shown it was possible for two people to mirror each other without a leader.

When I read an article he'd written in a science journal, I suddenly realized he was using principles from improvisational theater to make teams in his labs work better. Basic principles that work on the stage were working for him in the lab.

He struck a chord when he wrote about the pitfalls of assuming students are totally responsible for their own motivation, noting that "this can lead researchers to blame group members for their lack of motivation." Instead, he feels that it's up to the leader of the group to motivate the students, or else things can break down. This is the very thing that happens in the mirror exercise when the leader doesn't take responsibility for helping the other person to follow. They lose the connection that could keep them in sync. It happens when a teacher blames the student or a speaker blames the audience for not understanding what they have to say.

Comedians used to say to an audience that wasn't laughing, "These are the jokes, folks." And the audience might have been thinking, *Here's an idea: Try being funny.*

The responsibility really belongs to the person speaking, not the person listening. And now Uri was taking it a step further, suggesting that you can move a whole group to greater motivation. "The goal," he was saying, "is to provide people with the conditions that enhance their natural self-motivated behavior."

One of the ways he does this is to pay attention to their social interactions. The team meets once a week for two hours to discuss their work in the lab, but for the first half hour they don't talk about science at all. Instead, sitting in a circle, they talk about art, politics, sometimes just about how their week has gone. Only then does someone give a talk on his or her research, with the rest of the team often helping to brainstorm the project. He's found that the team's motivation rises along with this increased social connectedness.

This is very much like the approach we take at the Center for Communicating Science—bringing people together with an eye to social flow as a way to improve the flow of science. I felt I had to talk to Uri directly, so I put in a call to Israel.

For some reason, my phone company made it impossible to

dial the call directly but didn't tell me the price of having the operator place the call. I was smack in the middle of the joys of communication with a large corporation. Once the call went through, the cost of the operator's assistance, which I assume involved tapping buttons on a keypad, was somewhat higher than expected. Uri and I talked for a couple of hours, which cost about what you would pay for a small horse. But it was worth every minute.

Uri began his studies as a physicist and then turned to the field of systems biology. This is the field that uses mathematical models to figure out how a large number of component parts work together in complex biological systems—systems that range from molecules to entire species.

Uri has done important work in finding patterns of interaction in the complex workings of the living cell. The paper that brought him the greatest attention in this field was actually prompted by a moment of improvisation—something that was probably a natural outcome of his lifelong interests.

His life traced two paths at once. His mother was a physicist, and his parents had always wanted him to be a scientist. From the time he was born, they called him the Professor. But inside he also knew he was an artist. He decided, "Okay, I'll choose the science path and sometime when I'm old and gray, when I'm forty, I'll get back to art, to poetry." But art caught up with him before that. "In the army, we had these Purim parties, where we'd put on plays and make fun of the commanders. And I was always directing them and writing them and acting in them, and I found it exhilarating."

With a broken heart after the end of a love affair, he turned for solace to the guitar. His first taste of improvisation was improvising songs in bars. In graduate school, while he studied to be a scientist, he performed in plays, and one day when his director introduced him to theater improvisation, he realized he had found the art he was looking for. For the last fifteen years, he told me, he's been performing every Friday night with an improv group called Playback that specializes in acting out the life stories of members of the audience.

I had found a scientist-improviser.

For a long time, though, improv and science remained separate paths for Uri. "For years," he said, "I thought of myself as doing science from the neck up, and then I would go and do theater and feel that my whole body was alive. They were two different worlds, but slowly they started to connect."

Now he teaches improvisation to scientists and he's discovered a strong connection between the two worlds. "I'm very interested," he says, "in the parallel between going into the unknown in an improvisation and going into the unknown in science."

YES AND

Some of the most basic elements of improv have become useful to Uri. "Improv is listening. I listen now," he says. *Yes And,* the fundamental rule of improv, plays a large part in his work.

For improvisers, *Yes And* means you accept what the other player presents you with, without blocking it or denying it, and then you react constructively to it. You add to it. As an example, Uri says, "If one player says, 'Look at all that water down there,' and the other player completely blocks it by saying, 'That's not water, that's the stage,' then the scene is over. But if the player follows the principle of Yes And, he can accept what's been handed to him and add to it. 'Wow, what a lot of water. Let's jump in. Let's grab onto that whale.'" And they're off and swimming.

In the same way, Uri suggests that collaborators in a lab (or anywhere else) can listen to one another's ideas, no matter how odd they are, and add constructively to them. "A lot of blocking takes place in science," Uri told me, "but Yes And frees ideas to grow."

Once, after a run of disappointing dead-end calculations, a student said, "I wish we could just make a diagram of this on a piece of paper." Uri, instead of saying, "No, we've done that hundreds of times before," said, "*Yes, and* let's do it on a really *huge* piece of paper." Which they did, and the diagram they drew enabled them to see things they had never seen before.

Until then, the extremely complex system of how genes influenced one another in a cell had seemed like a hopeless mess to untangle. But Uri's saying yes to one more diagram (*and* a really large one) allowed him and his team to see for the first time that the genes interacted in three basic patterns that were repeated hundreds of times in the cell.

"So," Uri said, "the hopelessly complex network turned out to be simpler than anyone imagined." Not only was it a happy moment of collaboration in the lab, it was the breakthrough in Uri's work on systems biology that gained him attention in the field.

The same rule of acceptance and deep listening works not only among the members of his team but even in their individual observations of nature. Scientists try very hard to avoid imposing their biases on what they observe, and Uri told me it helps if a scientist can say "Yes And" to nature itself. "Instead of talking to the data," he says, "you listen to it."

All of this suggests to me that an inescapable product of improvisation is empathy. And that a combination of empathy and the more rational Theory of Mind is the very foundation of communication.

CHAPTER 9

Total Listening Starts with Where They Are

We've worked with thousands of scientists and doctors at the Center, and I've seen many times over that listening begins before you even start trying to communicate. You picture an audience and think, *What are they already aware of? Where should I start? How deep should I go? What are they actually eager to know? If I start too far in, will I be using concepts they don't really understand?*

Once, I came home from a visit to CERN, the European research organization headquartered in Switzerland, where scientists had just found the Higgs boson. All the newspapers were filled with stories of the discovery, and I was at dinner with a friend who was eager to hear what I knew about it. "What's the Higgs boson?" she asked. I gave it a try, even though I'm better at *asking* questions about complicated things than I am at answering them.

"Well," I said, "it's a particle that seems to give mass to all the other particles, and . . ."

"Wait a minute," she said. "What's a particle?"

I had started too far in. I had to back up. But there was even more to it than that.

You not only have to start with what they already understand, you have to know when to stop, or they'll feel swamped. If I had gone on into my sketchy understanding of how the Higgs particle is crucial to the Standard Model of physics, I'd not only have been in over my own head, I'd have been in over hers. She might have felt that crushing sense of hopelessness that tells you it's all too much, and she might never ask a question like that again.

I started again, earlier in the story, and muddled my way through a description of particles as the basic building blocks of matter, but I could see she still found it a little vague. Later, I asked Brian Greene, the physicist, how he would explain the notion of a particle. His description was homey. I wish I'd had it the night I'd come back from CERN. ("If you cut a loaf of bread in half, and then take one of the halves and cut that in half, and keep doing that, eventually you'll get down to the smallest bit possible. That's a particle.")

So, knowing what they're ready to hear is critical, and coming in too early or too late can be confusing. And, of course, having a clear, homey image like Brian's helps the listener visualize it and remember it.

Sometimes, though, they don't want an explanation at all.

If someone has a medical problem, do they want a detailed account of the facts (some do), or are they too vulnerable for that? Are they even able to *hear* the facts at that moment? Maybe they need presence more than knowledge.

Valeri Lantz-Gefroh, the director for improvisation at the Alda Center at Stony Brook, tells a story about this. Val is an accomplished actress with extraordinary teaching skills. I've watched her teaching workshops around the country. She's clearheaded and runs her improv classes with a kind but firm hand. She doesn't strike me as overly sentimental, but when she told me about an encounter between a doctor and a patient, she choked up.

"One of our medical students came and met me after one of our classes," she said. "It was a late class and he stayed until eight o'clock at night just to tell me how important improv has been to him. And the exercise that really stuck with him is the mirror exercise. He told me about a woman he had met a month earlier. She was dying. They had just found out that she had metastatic lung cancer. She had about two weeks to live. This student was on the rounds with the internist, and he listened as the internist shared the news with her. But he told her in such a way that, the student felt, she really didn't get it. She didn't seem to understand what was going on. The student said, 'I'd like to talk to her. Would that be all right?' The internist gave him permission and left.

"The student sat next to her and held her hand. He explained to her slowly and simply what was going on. He didn't use the word *metastatic*. He didn't use the word *malignancy*. Even those words felt like too much for her. And for the first time the woman cried. The student told me that . . ."

Valeri paused here for a moment and said, "*I'm* going to cry, just talking about it." After a moment, Val went on. "The med student said, 'I cried too. . . .' And then—for the first time—the woman began asking questions. And he was able to answer them. He said, 'It was the perfect mirror exercise. I had been leading, but then she took over and I followed. And, ultimately, what emerged was that I helped her understand death—and she helped me understand how to be a better doctor. It was exactly what the mirror exercise is about—that level of connection and active listening.'"

Our goal is not to make doctors cry, or to be overcome by the patients' emotion. We certainly don't want them to get trapped in "affective quicksand." In this case, though, the med student was reacting to a sudden ability to connect richly with another person. He was moved by his own personal breakthrough.

Even though he had a strong emotional reaction to his patient, he was engaged in a subtle exercise of leading her from one state of thinking to another—something that can happen in radically different situations. Even, it seems, in business.

LEADERSHIP

I'm certainly not learned about business, but I think there are similarities between what we've seen as we've worked with doctors and scientists and the experience of successful leaders.

Bill Gates, for instance, has written that when we assess how well suited people are to their jobs, it matters that we think about how well they relate to others. "Do they have both the IQ and EQ to succeed?"

His friend Warren Buffett, who eventually learned some of these same lessons, had a surprisingly slow start. He said in an interview that in school he was "terrified" to speak in public and avoided classes that he knew required it. Finally, realizing he would have to speak if he was going to succeed in business, he took a Dale Carnegie course. "I gave them a hundred bucks and I got my little diploma." It paid off in at least one way. While he was taking the course, he proposed to his wife. "So," he says, "I really got my money's worth there."

Aside from the homespun humor, what struck me most about Buffett's interview was that this once taciturn person could talk about communicating comfortably with 330,000 employees. He does it, in part, by getting right to the point. His messages to the

CEOs of his companies are models of economical writing. "I don't believe in two-hundred-page manuals," he said, "because, you put out a two-hundred-page manual, and everybody's looking for loopholes, basically." Instead, every two years, "I write 'em a very simple letter. A page and a half." In it, he reminds his managers that he and they all have plenty of money but not an ounce of reputation to spare. He cautions them to make business decisions that they wouldn't mind seeing reflected on the front pages of newspapers.

He speaks plainly and is aware of the readers' state of mind. He said he left a few things out of one of his reports to shareholders because "when I got to sixteen thousand words, I thought I might be losing them."

I'm fascinated by how CEOs manage large companies, mainly because my experience is so slim. When I've directed movies, I've never had to manage more than two hundred people at a time, and they all knew their jobs, so they pretty much managed themselves. In production meetings I'd tell them well in advance what I needed from each department, and then I'd leave them alone to go off and do what they knew how to do.

That's not to say I don't enjoy organizing things. On one shoot, for the movie *Sweet Liberty,* I had to reenact a battle scene from the Revolutionary War, with a hundred soldiers charging a line of another hundred soldiers. The sequence could have taken days to shoot, but we didn't have that kind of time. I had to do some fancy organizing. I asked seven stunt men to train a hundred pairs of men in seven different kinds of hand-to-hand combat and gave them a week to learn these different routines. Then I placed them in the scene so that no encounter would be next to one just like it. It had the look of a chaotic battle, but had been rehearsed in miniature.

We rolled seven cameras, and instead of taking days to shoot, we had a battle scene in two hours. For years, I was prouder of how I had organized that scene than I was of the scene itself. But I had done something that was mainly logistical. I hadn't needed to motivate people or rally them to a grand vision. And it was only two hundred people. It was a far cry from managing tens of thousands in a large corporation. How did CEOs do it on such a huge scale?

I was really curious, so when I met the man who headed one of these corporations, I asked him naïvely how he was able to manage 350,000 employees who were stationed all over the world. He regarded me with the kind of look you use if you're trying to decide if someone is actually crazy or just stupid. Finally, he said, "Well, I have good lieutenants." *Oh,* I thought, *so it's like how you manage an army.* And then I remembered my brief time in the infantry, where if you didn't do what you were told, you could be sent to peel potatoes, or locked up. Or shot. I wondered if you could actually make a profit with a business model like that. I was pretty sure there was more to it than that.

The experiences we were having in the workshops seemed to be pointing to the idea that while leadership is important, just as important is how leadership is communicated. On the one hand, you can command good performance from someone in exchange for not firing them. On the other hand, you might be able to ignite the *desire* in a person to perform well by tuning in to their state of mind. And, in fact, this has been shown by research to be the better way.

In 2001, Golnaz Sadri, Todd Weber, and William Gentry collected reports on 6,731 managers from their bosses and subordinates. They wanted to know how much it matters if leaders express an understanding of the emotions of those working for them. Apparently, it matters a lot. They found that "leaders who are rated by their subordinates as engaging in behaviors that signal empathic emotion are perceived as better performers by their bosses." In other words, around the time someone under you rates you as pretty empathic, your boss is beginning to notice that you're getting better results from your team.

One day I found myself talking about this at lunch with the CEO of a large public company. I felt he must be regarded as a pretty good manager since he's listed as one of the highest-paid CEOs in the country. I asked him what he thought about these two kinds of leadership: tough versus tuned in.

He told me he once worked under someone who was a *very* tough leader. As in, "Make a better presentation next time, or you're out." But in the company he runs now, he feels he gets better results if he starts by praising what they had been doing well and then urges them to bring their next presentation up to that level.

This is not just softening the blow; it's keeping in mind what the other person is thinking and feeling. And it's enlisting them in the effort to move them to their best. Instead of saying, "You've done a bad thing; don't do it again," he's saying, "You've done really good things; do more." The first gives them a vision of failure they somehow have to avoid, while the second gives them a model of success to live up to. The CEO I was having lunch with might not have realized it, but he was following the improv principle of Yes And. He was accepting what the other player was giving him and adding to it.

But, of course, it can't be faked. Faking praise is really lying. I've seen it fail badly between directors and actors, especially when the director tells you to do that wonderful thing you did two takes ago—and you know very well you did no such thing. It's too easy to spot the manipulation, and the person on the receiving end will probably be thinking, *This guy is aware of what I'm feeling, all right, but only so he can use it to make me do what he wants.* If we're actually able to get inside someone's head, it's not a good idea to be guilty of breaking and entering.

There are times, of course, when the only thing you can do is deliver the bad news.

Some of my limited experience as a boss has included the unpleasant task of firing people who, it suddenly turned out, were wrong for the job—that remarkable transformation where someone you thought was perfect has turned into a werewolf.

They haven't actually become something else, of course. They're the same perfectly fine people they were months earlier. But with all my supposed sensitivity and mind-reading ability, I hadn't picked up on who they really were when I hired them.

I could have avoided all the mind reading I needed in *firing* them humanely if I had been better at it when I was *hiring* them. Again, a deeper kind of listening would have helped.

More than once, I've found that in the first five or ten minutes of a job interview, the person applying for the position will tell me exactly who he or she is and what I can expect from them. They often don't realize they're laying it all out. Later, when I realize they really aren't right for the job, it's only then that I remember they had told me so. I just wasn't listening.

They don't frame it in a negative way, of course. Someone once proudly told me she was very, *very* good at paying attention to every single detail. It might have been a good idea for me to start asking myself a few questions, such as, *Is this someone who will turn out to be late with every task because the details are just too inviting?*

And it can be worse. As bad as getting the wrong person for the job is, getting the right person, but finding them incomprehensible, can be disastrous.

Someone once worked for me who handled a lot of my money. He seemed good at what he did, and the only problem was that when he talked to me I couldn't understand anything he said. He would go on at great length explaining *depletion deductions* and *tax-deferred exchanges* in complete double-talk. These terms started *out* in double-talk. How he made them worse I couldn't fathom. Somehow, I let this go on for a long time—well, actually for years—and then I couldn't stand it anymore.

"I'm sorry," I said, "we're going to have to part ways."

"Why?"

"Because we have a communication problem."

"What do you mean?"

"I mean I can't understand you."

"In what way?"

"Speaking. I can't understand you when you speak."

"Why is that?"

"Because we have a communication problem."

I was getting dizzy.

I was finally reduced to saying something not quite true, like "It's not you, it's me." And he asked me what *that* meant.

CHAPTER 10

Listening, from the Boardroom to the Bedroom

Almost all of us have some kind of communication problem that we don't know we have.

A large company realizes it's vulnerable to an attack by hackers. But the biggest hole in this company's firewall might be the gap in their own internal communication—a gap that often goes unnoticed.

Over the years, the management of the company has made minor, stopgap adjustments to their security code. But now the board of directors calls in their expert on information technology (the IT person) and asks for a full report.

She gives them the startling news that their network is wide open to attack. She tells them why, and what to do about it.

The board is listening intently for the bottom-line implications. Instead, the security officer carefully lays out the technical dimensions of the vulnerability. The board members don't really understand what she's talking about, so they decide to table the issue until the problem can be studied further.

Jay Leek, the chief information security officer at Blackstone Group, in an interview with the *New York Times,* identified the problem this way: Security people are steeped in technical knowledge—the knowledge they need to do their jobs—but they aren't tuning in to what the board members are listening for, or hearing. As Mr. Leek says, IT departments "have made what we do very complicated, very technical. We haven't been the most articulate communicating upward to a board . . . in such a way they can wrap their heads around the problem." He seems aware that good communication is the responsibility of the person delivering the information, not the person receiving it. Without understanding that, the sense of urgency can get lost in the weeds of technical expertise.

It's another example of how too much information can be a serious roadblock. At the Center for Communicating Science, we teach scientists that it's not necessary to tell the audience everything you know in one gulp. Sometimes, telling us just enough to make us want to know more is exactly the right amount. We gag on force-feeding. We're uncomfortable feeling like geese getting our livers fattened.

In the hacking problem above, the IT people weren't fully aware of the members of the board. They weren't speaking to what was actually in the board members' heads, but what was in their *own* heads—the purely technical dimensions of the problem.

Unless the people they're reporting to authorize improvements in the IT system, the work won't get done, and as Leek says, "It's people's nature where if they don't understand something, they tend to say, 'No.' " When saying no leads to a security breach, it doesn't hurt just the company, of course; it can hurt you and me when, later, our credit cards get hacked.

On the other hand, I was having lunch one day with a husband-and-wife team who help IT engineers communicate with businesses and government agencies, and they told me that sometimes it works the other way around. Sometimes they have to explain to the IT engineers what the *CEO* is trying to say. Maybe we should be grateful that anything at all gets done.

I take all this personally because I do happen to be an expert in one area of business, and that's as a customer.

As a customer, I get communications from corporations all day long. Almost everything I touch has advertising on it, but an even stronger message is delivered when I actually buy something. Every time I open a package, the company that made it is communicating with me. When I can't open a hard plastic "clamshell" container with scissors or a knife, or even a hammer, I wonder, *Has the president of the company ever personally tried to open this thing?*

I'm not alone in using a new product with bandaged fingers. A survey of two thousand people in England found that four out of ten people have been injured opening packages. In one case, the customer couldn't get into the box without cutting the plastic ties that held the new purchase in place—which was really hard to do because the new purchase was a pair of scissors.

Who's communicating with whom here? Retail chains want their suppliers to make it harder for thieves to open packages in stores. The manufacturers want their goods wrapped so securely they won't be damaged in international travel. Meanwhile, nobody seems to be thinking about my fingers. Will I be buying this thing a second time? Maybe not. If we could, 40 percent of us would choose, instead, to buy something similar, but wrapped in a package we can actually open.

But this is a side dish to the main course.

SELLING

The more obvious way business communicates with the rest of us, and with itself, is through selling. Selling, of course, is not confined to business. It probably predates business. There's a form of selling going on all the time among humans. Deciding on a family vacation is a sales job when one person's idea of the perfect vacation is backpacking through a majestic national park and another thinks it's careening through a wet theme park. But *selling* is not a dirty word, even though I once thought it was.

When I was young, I was hampered by the idea that selling was an exercise in which you bent someone to your will through manipulation and by not telling them what they needed to know to make an informed decision. I was uncomfortable with the whole process. This was not good, because as a young actor I was at one point trying to support my family by selling mutual funds. For quite a while, the only person I was able to sell any shares to was myself. I could hardly afford them, which was why I needed the job in the first place. But it seemed important to display my confidence in the fund. Still, loaded as I was with sincerity, I went into every meeting with the idea that the prospect had something I needed (money) and I had a sales pitch I could use to pry it out of them. This was getting me nowhere. While I was speaking to my prospect, I was thinking about what I had to *tell* them and not what they were hearing, thinking, and feeling. I was focused on myself, not on them.

After an agonizing series of failures, I had a sudden awakening one day when someone came to me without my seeking her out. She was a masseuse who had a few thousand dollars she wanted to put into a good investment and was interested in hearing about my mutual fund. For the first time, I saw myself as helping someone. Now *she* was the one who needed something, not me. Suddenly, I was engaged by what *she* was feeling and thinking. I'm sure most good salespeople know their job is to help the other person get what they want, and even better, what they need, but for me it was a revelation.

SOCIAL AWARENESS AND THE EMPATHIC STRATEGY

When things go right in selling, the salesperson is using *social awareness,* a term that was first used by the Columbia University psychologist Edward Thorndike in the 1920s. Social awareness became one of the key attributes of *emotional intelligence* as described in the work of Daniel Goleman. Goleman has refined the notion of social awareness to three separate steps: first, having an instantaneous, primal awareness of another's inner state (empathy); then, grasping their feelings and thoughts (Theory of Mind); and, finally, understanding—or, as he says, "*getting* complicated social situations." Goleman has described social awareness as "the ability to identify a client's or customer's often unstated needs and concerns and then match them to products or services." He adds that "this empathic strategy distinguishes star sales performers from average ones."

In my own experience, and in the research I've read, empathy also seems to lead to a greater ability to be patient. If so, it probably plays a part in enabling a successful salesperson to take "the longer view" that Goleman says is valuable to both sides of a sale. As he says, social awareness

also means taking a long-term perspective, sometimes trading off immediate gains in order to preserve customer relationships. A study of an office supply and equipment vendor indicated that the most successful members of the sales team were able to combine taking the customer's viewpoint and showing appropriate assertiveness in order to steer the customer toward a choice that satisfied both the customer's and the vendor's needs.

Interestingly, Goleman speaks positively of assertiveness. Not assertiveness aimed at getting what *we* want, but assertiveness in providing what the *customer* really needs. And isn't this what happens in a loving relationship when one partner helps steer the other toward his or her best instinct and values, regardless of the short-term benefits to either of them? (As in, "Are you taking that job because it pays more, or is it really what you want to do with your life?")

Very much like in a close personal relationship, success in selling has been shown in a number of studies to improve when the salesperson listens to the client and actively expresses an understanding of their thoughts and emotions. It leads to more trust and higher sales, and I would guess more satisfaction on both sides. For some people, this comes naturally, and for others it's eventually discovered. But somehow, it doesn't happen often enough.

Why? If there's a connection between deeper listening and the bottom line, you'd think the connection would be self-reinforcing. But for most of us, it takes extra effort to pay deliberate attention to what the other person is thinking. Even if it's the one person we're most devoted to. In fact, my guess is that good communication with the person you're closest to may be the hardest challenge of all.

COUPLES: *REALLY* ACTIVE LISTENING

He leaves his socks on the floor; she can never find her purse. He always tells jokes, but forgets the punch lines. In the shallow waters of such petty annoyances do great marriages run aground. These little irritations tend to mount up, but if what I'm hearing from researchers is true, then a richer kind of listening can produce a little cooperation and a lot less friction. For instance:

When men share the housework with their wives, it leads to more sex. No, really. It does.

Dan Carlson and his colleagues at Georgia State University did a study in 2014 and found that "couples who shared routine housework equitably had the most sex, about 7.74 times a month." And those couples achieved the highest level of satisfaction, too.

But even if men believe this to be true, how motivated are they to act on this theory? In the United States, and across the world, only about 30 percent of men who live with women share the housework equally.

Why? Many men have known for a while now that doing the dishes can be foreplay. But it's not so easy to remember the connection between scrubbing a pot and connubial ecstasy when all you're facing is the pot.

What if empathy is brought into play? Research suggests that people cooperate more as empathy increases.

When a husband looks at the dishes in the sink and thinks, *I guess I ought to do something about that,* chances are iffy that he'll act on that impulse. But if he looks at the pile in the sink and imagines his partner seeing it in the morning—in other words, if he sees it through her eyes—he'll be more aware of what she'll feel, and that brief moment of empathy could engender some cooperation on his part. And then, who knows? Washing the dishes might seem easier to do, it might feel more authentic, and it might lead to something interesting later in the evening. At least, it couldn't hurt.

There's another great cooperation killer, the Sound of Certainty: the triumphant, but self-defeating, tone of voice that announces, *I know what I'm talking about and that ends the discussion*. It's a tone that doesn't invite the other person in, but, instead, diminishes them to the rank of outsider. Sometimes, this tone is a sound we make without even realizing it. We might not be aware that we've just closed the door on the other person until we see it register on their face (if we're actually paying attention to their face). The tone of what we say can be deadlier than the words themselves. It can change a perfectly well-intended phrase into something disastrous, and it's not largely under conscious control.

I've read a lot of tips on communication for couples. And many of them are useful, like *active listening* (letting the other person know you've heard what they said) and making "I" statements (concentrating on how the other person's behavior makes you feel, rather than simply denouncing their bad behavior). The problem is that some of these tips can have the opposite effect if you're not aware of what's going on in the other person's head—or in your own. Or if you're not aware of how your own feelings are affecting your tone of voice.

Active listening can easily morph into veiled hostility. There's a difference between "Let me see if I heard you correctly," and "Let me see if I heard you *correctly*!" (with a stern face).

If you were just collecting tips, a good tip would be to remember not to listen actively in a hostile way. But it's difficult to tiptoe our way consciously through the hostility patch on the way to genuine connection. We need to get there effortlessly, naturally, because the more conscious we are of producing a neutral tone, the more manufactured it's liable to sound. I know from my experience as an actor that an authentic tone of voice is produced deep inside the brain, not in the voice box. My layman's guess is that an authentic tone, a tone that means what it says, relies heavily on the social circuits in the brain. This kind of tone is produced less by a decision to sound a certain way and more by our relationship with the other person.

I've seen this happen over and over in improv classes. When people manage to tune in to their partners, there's an effect on muscles in the throat and face that are not under their conscious control. The tone of voice changes and the expression on the face softens, not because they decided to, but because that's what happens when you connect and make contact with someone.

You can hear the difference in people who have been trained to make social contact more mechanically.

For instance, there's that eerie moment when I've checked into a hotel almost anywhere in the country and realize that room clerks everywhere have been instructed to behave in a welcoming way. It's eerie because they all say the same thing in the same disembodied tone of voice: "How was your flight? How's your day been so far?" Usually, my day has been okay until I hear the robotic sound of their scripted questions, a sound that's less than welcoming. I would think that training clerks to make real contact with guests would be more fruitful than telling them to use a specific set of words and that all-purpose cheerful tone—a dry formula. It shouldn't be that hard to do.

Checking in is not a complicated interpersonal action. It's not the subtle exchange that couples go through, filled with hidden meanings and subterranean messages involving life-changing decisions—like loading the dishwasher.

Filling up the dishwasher is, apparently, a loaded activity for couples. Bosch, a company that makes these machines, has published a study in which they found that 40 percent of us fight about how we load the dishwasher. Sixty-one percent of those fights are about whether or not to prerinse the dishes. This seems like an excessive number of people fighting about rinsing dishes, but that's what Bosch claims. Thirty-nine percent are arguing about whether forks and knives should point up or down. This I can understand; it's surprising that more people don't die by falling on forks in the dishwasher. On the other hand, thirty percent argue about whether to put plastic containers on the top rack. These people should get a hobby.

I don't know if this was a rigorous study, but it *is* a reminder that everyday events in our lives can be loaded with volcanic emotion and tectonic slippage.

Couples might have an especially hard time with communication because so much of it is nonverbal. We might not think our true feelings are showing, but our partners can read our feelings nonetheless. All the more reason to be good at reading our own emotions, as well as the emotions of the other person, and then being able to regulate those emotions. Or the volcano could blow.

And it isn't easy. We have to practice until it becomes second nature, and even then, it can feel like a burden if only one person is doing it. I was talking to a friend who had seen the power of the improvising mantra "Yes And" in an improv class and thought she would try it at home with her husband.

"I did it for two weeks," she told me. "It worked. I finally quit."

"You quit? Why did you quit?"

"Because I couldn't stand doing it anymore."

It's probably better if both people do it.

LESSONS FROM M*A*S*H AND MUSIC

I learned something during the eleven years we were filming *M*A*S*H* that changed for the rest of my life the way I prepare to go in front of an audience. When we began the show, we knew that somehow we had to become a tight-knit group. The characters in our stories lived together in harrowing conditions of overwork, extreme cold, extreme heat, and the barbarity of war. They knew one another better than any family. We had to create that sense of intimacy, and yet the actors had never met before. I don't think any of us consciously devised a strategy to bring this sense of closeness about, but we gravitated to a solution that was utterly simple. And it transformed us.

During a film shoot, there are long stretches while the set is being lit. Actors will often spend that time in their dressing rooms, apart from the other actors. They'll meet for a brief moment before the scene is shot, rehearse once or twice, and then the camera rolls and it's opening night.

But it was different for us. We would spend most of our time between shots sitting together with our chairs in a circle, making one another laugh. We would occasionally do something useful like going over our lines together, but I think the majority of our time was spent laughing. When they called us to the set, we kept the connection going as we walked across the sound stage. When the camera started rolling, we were still engaged, but now we were using the words and emotions called for in the script. That connection gave us a jolt of life. It was by far the best preparation for acting I had ever come across—for me, far better than sitting alone, summoning up emotion and memory.

I need the contact with the other players to bring me out of myself and into the spontaneous moment. Every time I've done a stage play, I've tried to get the other actors to sit together for about an hour before every performance. Within a few minutes, there's a general air of hilarity. You're vulnerable when you laugh. You let the other person in. And when you get on stage, the channel between you is open, and whatever they do registers on you. You're like a spider sitting on a web that responds to the slightest tickle of motion. But the secret, I think, is not so much the laughing; it's the connection.

Even performances of music seem to be enhanced by establishing that connection.

I was having lunch one day with the clarinetists Stanley Drucker and his wife, Naomi Drucker. Stanley spent almost sixty-three years with the New York Philharmonic, starting with the orchestra as its youngest member and retiring as its oldest, and Naomi is a founder of the American Chamber Ensemble, and has often played with the Philharmonic.

"Do you ever pay any attention to the audience when you're performing?" I asked them.

Stanley nodded. "It helps a lot." I was a little surprised he noticed them at all.

"How are you aware of them?"

"Seeing their faces. If they look attentive, it helps. Sometimes it's a challenge. I see them really paying attention and I want to say, *You think* that's *good? Watch what I can do.*"

We laughed, and Naomi said, "It's really nice sometimes when you catch their eye, and they look at you and smile."

So it seems that, like actors, musicians can be in a dynamic relationship with the audience. They certainly are with their fellow players.

When I see the great Yo-Yo Ma come on stage for a concert, no matter the gravity of the work he's about to perform, he stops to greet other members of the orchestra. He waves at them, hugs them, jokes with them. Then he sits down, closes his eyes for a moment, and tears into the music, connected both to the music and to the players around him.

Yo-Yo pays the same attention to what the audience is receiving as our scientists do. In an interview in *Listen* magazine,

Yo-Yo said, "You have to keep both things in mind at the same time: What's the story? Where are you in the story? . . . And what ultimately makes it memorable is that the thing that I care so much about lives inside somebody else's head, and that it's *received*." He asks himself the same kind of questions we suggest our scientists ask: "Who is receiving it? Who are you? Why are you listening to this? Why would you care? *Should* you care? . . . I think," Yo-Yo says, "that that is an unbelievably important and probably less explored part of music."

I saw this in action with Toby and Itzhak Perlman at the Perlman Music Program, a summer music school for gifted young musicians from all over the world. The kids had been telling Toby they were uncomfortable talking to the audience when they announced a piece before performing it. Toby invited me to teach a class to see if the work we did with scientists might be helpful to the young musicians. I was excited to try it, because I was pretty sure even a few hours of improvising could make standing in front of an audience a little easier. I also had a secret wish to see if improvising would have any effect on their music.

We set up in a large dining room at the camp. There were about twenty young musicians and a group of faculty and students who were curious to see the workshop and were just daunting enough as an audience for the students to practice on.

I had asked the students to bring their instruments and started by having them speak for a minute about a piece of music and then play the first thirty-two bars. Some were severely shy, especially a Korean girl who looked about thirteen, but she didn't back down, and when it came time to play the improv games, she pitched in.

I asked them to make imaginary objects out of space and pass them around in a circle, just as I would if I were working with scientists. Next, they tossed balls made of nothing to one another, but the kids had to make sure the ball they caught was the same size and weight as the one that was thrown to them. As always, the better they got at that, the more you believed you could see the ball passing from one person to another. The kids were delighted to see imaginary objects coming into existence simply because they were observing one another and responding—making contact.

Then I asked them to toss not a ball, but an emotion around the circle. That was hard to grasp at first. It was confusing enough to toss around a ball that didn't really exist; now I was asking them to pass around a *feeling*. To get them started, I made a sound and a gesture that I hoped conveyed a sense of joy and tossed it to the girl standing next to me. Her job was to catch it, the same way she had caught the imaginary ball, then mimic the emotion and pass it on to someone else. The emotion made its way around the circle, and soon the natural expressiveness of musicians took over and they were tossing passions to one another.

But next, before they passed the emotion on, they had to let it morph into another emotion. They would now receive an emotion, copy it, allow it to turn into a completely different feeling—and then toss that to the next person. No one knew what to expect from the player next to them, but they would catch it, become it, let it change into something else, and pass it on. The grief they received, and responded to by acting it out, could turn into rage, which, tossed to the next player, might become uncontrollable laughter.

This process of allowing something you receive from another person to transform into something else is one of the most interesting experiences in improvising. It's a delightful moment of creativity. You're manipulating an imaginary object or acting out an emotion, and suddenly you see it changing into something else. You don't consciously put the new thing there; you accept what the other person gives you and then, as you lend your body to it, you see it becoming something else.

In a way, as the young musicians did this, they were practicing Yes And. You accept what you get from the other person, and then let it grow into something else. You keep moving things forward. You don't cut off the flow. You build on what the other person offers you. You give them a green light instead of a stop sign.

Then came the game I had been waiting to try on them, one I had never been able to do with scientists. I asked them to pick up their instruments and form a circle. This time, they were to make an emotional sound with their *instrument*—the idea wasn't to play an emotional tune, just to try to make the *sound* of an emotion—using their violin, viola, or cello to make a kind of abstraction of emotion, even if it was an ugly sound.

Some couldn't help playing snatches from the string literature that were associated with certain emotions, but I would bring them back to just making sounds.

Soon they were receiving an emotion from the person next to them, letting it morph, and then using their instrument to toss a new emotion to the next person: the sound of anger, turning into the sound of love, becoming the sound of disgust. I wanted to see if the sense of transformation and spontaneity they had been allowing to happen using their bodies would somehow show up later in their playing if they tossed emotions using their instruments.

But first, we played word games, because after all, I had been asked to help them with speaking to an audience when they were onstage announcing the piece they would play.

I asked them to sell something in gibberish while the rest of the class tried to figure out what the product was, just picking up clues from the pitchman's body language and tone of voice.

In another game, they would sit in a chair and try to figure out what their emotional relationship was to someone—just from reading clues in the way that person told them about their favorite book or movie.

I noticed that the shyest one there, the Korean girl, was starting to leap up and join in every time she had the chance.

After three hours, they faced the audience again. They announced their piece of music and played the first thirty-two bars. The Korean girl was beaming. This time, she seemed glad to face the audience. In fact, all of them were more at ease and sometimes even playful introducing their pieces. But the shock for me—and it was a happy shock—was that almost all the kids seemed to be playing the music with more freedom, even more joy. I wanted to make sure I wasn't hearing things. I turned to Itzhak.

"Am I kidding myself, or did they actually play a little better?"

"They did," he said. "Except for him . . ." He pointed to one of the students. "He was amazing to start with."

That's okay. I was happy with nineteen out of twenty.

IMPROV ALL AROUND

Life, of course, is an improvisation. You don't know what's coming next. Your partners on the stage of life will always say something that throws you a curve. "I'm not happy anymore. I need some space." . . . "We're thinking of giving your job to your assistant." . . . "I'm not going to college. I'm getting a tattoo and going to Africa."

What do you say? "No, this isn't possible; it's not going to happen?" It's not only possible, it's what is.

As hard as it is to keep the conversation going when your child wants to drop out of college, an even harder thing to react to is when someone demands something while pointing a gun at you. This happened to my friend Larry Gelbart. Larry brilliantly wrote most of the first four years of *M*A*S*H* and was one of the most creative and empathic people I've known, but one act of empathy stands out.

Larry was coming home late one night to his house in Beverly Hills when someone stepped out of the bushes and leveled a gun at him. The man wanted money, jewels, whatever was in the house. He told Larry to unlock the door.

Larry let him into the house. Then what did he say? *You can't do this? How dare you? This is an outrage?*

He looked at the young black man, a kid really, and said, "You don't need to do this. You're too smart. What if I help you get a job?"

The conversation probably took longer than that, but I remember it this way because it startled me so much when Larry told me.

The kid put away the gun and the next day Larry got him a job.

Yes And.

But a story like this raises a question. Was Larry just lucky that he had this ability to connect with the young man, to sense what he was feeling? What if Larry didn't have an abundant, natural supply of empathy? Are we stuck with whatever measure of empathy we come into the world with?

In our work with scientists, not only is one of our core assumptions that a deep awareness of the other person is at the heart of good communication, but we also believe that empathy can be increased.

But what if empathy is something you either have or you

don't? Are we born with a capacity for empathy that remains pretty much the same throughout life, or can it be enriched and deepened? Can people be trained to have more empathy?

When I met with Helen Riess, she told me how she had realized the importance of empathy in her own work when she and her patient were both hooked up to a machine monitoring their emotions. And then she told me how that prompted her to train *other* doctors to be more empathic with their patients.

I had two big questions: How does she go about it?

And does it really work?

CHAPTER 11

Training Doctors to Have More Empathy

Once Helen Riess realized she had been missing her patient's emotional ups and downs, the experience changed the way she related to patients, and it changed the way she trained other doctors to care for *their* patients.

The first thing she had to do, she told me, was help doctors understand that in fact it is actually possible to learn to become more empathic. She maintains that it's not something you either have or you don't. Rather, most of us come equipped with the mental hardware for it.

She told me, "I introduce them to the neuroscience of empathy"—in other words, she teaches them how our brains are wired to receive the thoughts and feelings of others. Then she caught me by surprise with this: "Nowhere in medical school is this taught—the whole concept that there *is* Theory of Mind— that we can do a kind of mental time travel and enter the mind of others."

For therapists, entering the mind of the patient is regarded as central to treatment. Helen points out that the psychoanalytic theorist Heinz Kohut said that without empathy there *is* no treatment. There are countless papers in the literature cautioning therapists about the need for empathy. But for the medical doctors Helen was training, it was a novel idea that this was a tool of the trade. If they still carried little black doctor bags, empathy was unlikely to be in them.

AFFECTIVE RESONANCE— RESONATING EMOTIONALLY

The key to putting that empathy tool in the bag is making use of the brain's ability to reflect what another person is going through. As Helen said, "There's a temporary sharing—mapping—on the observer's brain. That temporary sharing has helped us through evolution to survive, and it's necessary for affective resonance."

I had a little trouble understanding what *affective resonance* is. She explained that it's the feeling of connectedness we're able to get with other people; for instance, "from looking people in the eye, from unconsciously having our bodies fall into sync with theirs. When we're tuned in to someone," she said, "these mirror functions happen almost automatically." And in a way, we can have the sensation of sharing their emotions—resonating emotionally.

In her class, Helen introduces doctors to this notion by having each of them tell a story to the person sitting next to them. Once they're engaged, she brings the awareness of their connection to the surface. "Did the person you told the story to seem interested? How did you know?" The doctors usually say they could tell because the other person was nodding their head,

ALAN ALDA

making eye contact, laughing at their jokes. They have the experience of being listened to, and they become aware that they can
listen more deeply to the person listening to them. This exercise
reminded me of our version of the What's the Relationship?
game we do in improv sessions, and that I described in Chapter
5, where a scientist explains her work to someone who's trying
to guess his relationship to her just from the way she behaves.
The focus is on connection—on reading the other player.

THE PATIENT'S PERSPECTIVE

When she works with doctors on how they talk with their patients, Helen Riess suggests something that sounded, to me, a
little unusual. She recommends getting the patient's perspective—by asking the patient, "What do *you* think is causing the
problem?"

This seemed like an odd thing to do. But apparently it's been
going on for some time among physicians who feel it leads to a
deeper kind of contact with their patients. A few weeks after I
met with Helen Riess, I was talking with a pediatrician who told
me a story her professor had told her when she was in med
school.

"He was teaching us how to listen to patients," she said. "He
told us how when he was a young doctor there was a very difficult case that none of the residents could figure out. He was sort
of the junior person on the team, and he thought, *I'm not going
to be able to figure it out if all my supervisors can't figure it out.
So I'm just going to talk to the patient and ask him what he
thinks is going on.* So he went to the station and said to the patient, 'You know, no one can figure out what's wrong with you.
What do *you* think is wrong with you?' And the patient, who
had been turning his symptoms over in his mind, said, 'Doc, I

92

think I have malaria.'" For some reason, this had never occurred to the doctors examining him.

The pediatrician telling me the story smiled. "And so they did the tests for malaria, and he had malaria."

Getting the patient's perspective is what Helen Riess calls *cognitive empathy:* an important first step in the doctor's ability to resonate emotionally with the patient. That emotional resonance, she says, helps us know whether the other person is at ease with us. Our sense of what they're feeling is based on what we're feeling, because, as Riess says, "most connections are mutual."

SELF-REGULATION

Whatever brain circuits are at work at this point producing this sense of mirroring, the doctor has to avoid getting swamped by the patient's emotion and sinking into "affective quicksand." Self-regulation has to take place: You know what the other person is going through because you recognize their emotion in yourself, but you don't have to act out that emotion. You take responsibility for regulating your own feelings.

EVEN MORE EMPATHY TRAINING

As I read more research, I saw that Helen Riess wasn't the only scientist teaching empathy with techniques similar to those of improvisation. At Boston College, Thalia Goldstein and Ellen Winner wanted to know if students would become more empathic if they were given acting training. Since actors have to step into the shoes of another person, the researchers wondered if experience in acting would lead to growth in empathy and Theory of Mind.

To test this hypothesis, they did two studies, one on elementary school students and the other on high school freshmen. All the students were given standard empathy and Theory of Mind tests before and after the training, to see if the training had any effect.

It did.

Both groups of students trained in acting showed significant gains in their empathy scores. Adolescents showed even more progress than younger kids: They had significant gains not only in empathy, but in a test of Theory of Mind, as well. Control groups that had been given other kinds of arts training, such as training in music or visual arts, showed no such improvement. Only theater training did it.

Other studies have pointed to similar results. Training in improv and acting in general both lead to improvement on standardized tests designed to measure a person's ability to connect with the state of mind of another person.

But does this work in real life? That was the question I was most interested in. When we train doctors to have greater empathy, do they not only do better on tests, but do their patients actually get better? Will it pay off in better health for the patients?

The answer is yes.

A meta-analysis of 127 studies on how patients react to their doctors' communication skills found that "expressing empathy and concern, providing information on the illness and its treatment, and encouraging patient participation in decision-making" made a real difference. The study found that those patients were "19 percent more likely to follow their physicians' recommendations on medication, diet, exercise, and screening than those patients whose physicians communicated poorly."

And there were demonstrable improvements in the health of the patients. In studies with diabetics, for instance, when the patients rate their doctor as empathic, their cholesterol and blood sugar numbers improve. Flu patients get better sooner, and even the common cold doesn't last as long and doesn't feel as bad.

Again, as in other research, gender plays a part. In a meta-analysis of twenty-three studies, female physicians visited with their patients 10 percent longer than male physicians did, and were more likely to ask questions about how their patients were feeling emotionally. The behavior of the *patients* of women physicians was more interactive. These patients spoke more, and disclosed more medical information to their doctors.

As in other studies, these results probably have something to do with the observation that women tend to be better at picking up nonverbal cues. But across the board, regardless of gender, doctors who display empathy get better results.

Even *doctors themselves* are said to benefit. One study has indicated that when senior medical people show empathy, there's a kind of "emotional contagion" in the hospital and *everybody* feels better.

I know. It sounds like it cures everything. But empathy does have a strong effect, even on malpractice suits. It reduces them appreciably.

The letter I received from the dentist who gave me a smile-ectomy is still vivid to me. Every sentence expressed his concern, not about how I felt, but about whether or not I would sue. And that wasn't an unusual stance of self-protection. For years, lawyers often advised doctors to "deny and defend" whenever the subject of malpractice came up. But the advice didn't seem to help much. The number of malpractice suits stayed high.

Lawyers may still be offering doctors that advice, but some hospitals are taking a different approach, one based on better communication. A few years ago, the University of Michigan began a program of encouraging doctors who had made a medical mistake to talk about it with their patient—admitting the error and even apologizing. From the lawyers' point of view, expressing regret was a mistake in itself, but within six years, the number of claims and lawsuits dropped from 262 to 83.

EVONNE KAPLAN-LISS

There are dozens of reports of advantages that flow from this kind of personal connection—too many to ignore. But you can't just tell people to be more empathic. It takes a skilled communicator to teach it—like Dr. Kaplan-Liss.

For a young woman, Evonne Kaplan-Liss has had an extraordinary life. She was a child actress with a small part in the movie *Annie Hall*. At about nine, she auditioned to play the lead role in the Broadway musical *Annie,* and came in second to Andrea McCardle. In her teenage years, she needed medical attention and her doctors recommended an operation that was "state of the art." She and her parents agreed to the procedure, assuming *state of the art* meant "best possible." What they didn't know was that, instead, the doctors were using the term to describe a procedure that had hardly ever been tried on children. "State of the art" just meant *relatively new.*

It was a learning experience in poor communication that led to twenty-one surgeries over the next thirty years.

Evonne studied journalism and became a journalist for a while, then switched to medicine and became a pediatrician. She now trains medical professionals across the country. Using her experience as an actor, journalist, doctor, and even a patient on the operating table, Evonne is a formidable speaker. After one of her talks, an anesthesiologist who works with cataract patients wrote her and said she had changed her practice with just one piece of advice from Evonne: "Don't start off with all the details. Get to the bottom line. 'You're going to be fine. I'm going to make you comfortable. You're not going to be in any pain.'

"Previously," the anesthesiologist wrote, "I would ask the patient if they 'did okay with anesthesia' and would immediately proceed into what they could expect. I felt like I was doing well because of the detail I was providing; who else spent the time doing that? (Aren't I wonderful . . .)

"Now," she wrote, "I start the speech with, 'I'm your anesthesia doctor and I'm here to keep you safe and comfortable.' I swear to God, the patients visibly relax when I tell them this—and only *then* do I discuss the specifics."

This anesthesiologist has also asked the whole OR team to change the way they speak to patients. Evonne read to me from her letter: "For example, instead of saying 'I'm going to drape your face,' now we say, 'I'll be placing this light cover over your eye to protect it.'"

Evonne looked up at me, her eyes flashing: "Right? What would you rather hear? 'I'm going to protect your eye,' or 'I'm going to put a *drape* over your head?' Which is what I heard twenty-one times."

EXPERIENCE, NOT WORDS

Once in a while, as with this anesthesiologist, a word of advice can have a profound effect, but Evonne doesn't rely on words alone. She and the rest of our staff have developed games and exercises that take med students step-by-step through *experiences*. Words can introduce you to an idea, but we think it takes an experience to transform you. It's like the venerable and sage advice to writers: show, don't tell. But it goes even further. Don't just hear about; *do it*—and get transformed.

TIPS

Even though I don't much like them, I have to admit that tips can sometimes be useful. Here are a few that have been good to me.

The Three Rules of Three

1. When I talk to an audience, I try to make no more than three points. (They can't remember more than three, and neither can I.) In fact, restricting myself to one big point is even better. But three is the limit.
2. I try to explain difficult ideas three different ways. Some people can't understand something the first couple of ways I say it, but *can* if I say it another way. This lets them triangulate their way to understanding.
3. I try to find a subtle way to make an important point three times. It sticks a little better.

But even though I've discovered a few tips that have helped, for my money, tips tend to be anemic when they don't come fortified with experience, or with a vivid story that lets you enjoy a vicarious experience.

I was once asked to write a list of tips on how to communicate well, and I resisted. I finally hammered out three, but they were so snarky I never sent them in:

1. Beware of tips. Tips are intellectual and often mechanical. They don't transform you. An experience transforms you. There's a stretch of road I've driven down

many times where I used to ignore the speed limit sign. One afternoon, I got a speeding ticket and I never ignored the speed limit again. The sign was a tip. The ticket was the experience.

2. Make a personal connection with your audience. Look them in the eye and speak to them as if they're a close friend and not a multitude. This is, of course, impossible to do just by reading this tip. Experience is what transforms you. (See Tip 1.)

3. If you can, experience improvisation. It will focus you on the other person. Improv games allow most of the tips about public speaking to become second nature, rather than forced and mechanical. A common tip advises you to vary the pace of your talk. Improv puts you so in touch with the audience that varying your pace happens automatically. You do it without thinking, which is the only way it will be effective. Trying to follow tips, rather than letting behavior emerge from the experience of improvisation, can actually make a talk more wooden. You see it in the pauses speakers take when they try to apply the tip to pause every few sentences. During these mechanical pauses, the air is dead. But when an improviser takes a pause, something meaningful is happening. She pauses because she's watching the audience to see if they understand her, and she's actually thinking of what she ought to say next. The pause is filled with something that's happening between her and the audience. She's alive.

Not everyone, of course, has a chance to join an improvisation class or take a course in mind-reading skills. So I wondered if it might be possible to do this kind of work on one's own.

Maybe I could shore up my own feelings of connectedness when they started to lag.

I asked myself an innocent question: Are there solitary exercises that might work? I should have known better than to ask. I have a history with this kind of thing. I soon found myself spiraling through a set of peculiar adventures.

PART TWO

Getting Better at Reading Others

CHAPTER 12

My Life
As a Lab Rat

TESTING AN EMPATHY EXERCISE

I have a habit of experimenting on myself.

In my twenties, I was fascinated by the notion that a person's temperature goes up and down during the day. So, to test the idea, for several months I carried a thermometer in my pocket and took my temperature every hour. No matter where I was. Understandably, I appeared a little weird to the people I had meetings with while I had this thing sticking out of my mouth.

I got caught up in the same kind of mania when I started looking for ways to practice mind reading on my own. I wanted to see if I could improve on my abilities at empathy and Theory of Mind, and I was searching for a kind of personal human-contact workout gym.

I started by practicing reading the faces of strangers—people in the street, store clerks, taxi drivers—trying to get inside their heads and figure out why they were saying what they said, the meaning of their body language and tone of voice.

I practiced *listening* to people, asking their opinion about things. Even in casual encounters, I tried to see things through their eyes.

I did it everywhere I went. It was a little less obvious than walking around with a thermometer in my mouth, but no less obsessive. Surprisingly, it seemed to be having an effect on me. Maybe it was causing a change in the tone of my own voice or the look on my face. *Something* seemed to be changing, because the behavior of other people was becoming different.

One day, I hailed a taxi at Columbus Circle. The cab pulled up and the driver rolled down the passenger window and called out to me, "Where are you going?" When drivers ask you this before you get in the cab, it means they won't take the fare unless they like where you're going. This is against the law. I drove a cab for a while in my twenties and I know how annoying it can be to have to drive to far-flung places—I once had to dig my cab out of a snowbank in the Bronx at two in the morning—but I went where the passengers wanted to go, because I knew I had to. When I get asked this question now, my usual response is not to identify compassionately with the driver, but to stoke the fire under my boiling blood. *I went, pal, and so can you!* is roughly my thought, and I walk away without negotiating.

But this time, I looked him in the eye. I saw no hostility. *It's the end of his shift,* I thought. *He wants to get home.* Suddenly, I was all empathized up. I gave him the address, and he let me get in the car. I was surprised I didn't feel my usual resentment at having to audition for a cab ride, but then he said, "What's the cross street?" This was another flash point. *I've never been there before,* I thought. *How am I supposed to know the cross street? Isn't that sort of* your *job?* Ordinarily, I would start boiling again. Instead, I took out my iPhone and opened a map. "I'm looking it up for you," I said. We were getting to be real team-mates.

"Thank you," he said. "I'm trying to get to a bathroom. I've needed to go for the last half hour."

"So, look," I said, "just drop me at Eighty-sixth and Broad-way. I'll walk the rest of the way."

"No, no," he said. "You're a kind person. People get in this cab, they don't care about other people. I'm taking you where you're going."

"No, look," I say, "it's all right. It's only a couple of blocks."

Now we were in an ecstasy of cooperation.

"Don't make the turn here," I say, "you'll have to go four blocks out of your way. You'll waste five minutes."

"No! You're a nice person. I'm taking you to the door."

I couldn't stop him. This man was sacrificing his bladder for me. I wished I'd never started the whole thing.

I stopped practicing empathy for a while; it was exhausting. But I couldn't stay away for long. I started in again, with a slight shift. I began to look at people's faces not only to guess what they were feeling, but to actually name it. I would mentally at-tach a word to what I thought was their emotion. Labeling it meant that I wasn't just observing them; I was making a con-scious effort to settle on the exact word that described what I saw. This had an interesting effect on me. First, I felt I was listen-ing more intently to what they were saying, even if earlier I had found them somewhat boring. And secondly, I would feel a sense of comfort, almost a sense of peace, come over me. It seemed a little bizarre, but so far it wasn't causing people to sacrifice their organs for me.

The feeling of peace was probably just a sense of relaxation. Whatever it was, naming other people's emotions seemed to help me focus on them more and it made talking to them more pleasant. I had no idea, of course, if other people who tried this would have the same experience, or if it was true that I was building up some empathy. Someone would have to do a study on it to find out. But I didn't expect anyone to devote research time to studying such a cockeyed idea. On the other hand . . .

NAMING EMOTIONS AS A WAY TO INCREASE EMPATHY?

In a conversation with Matt Lerner, the scientist who taught improv to kids on the autistic spectrum, I told him I had been trying to find a way to get the same benefit without the person's having to join an improv class. Some kind of solitary exercise. I told him about how I was silently labeling other people's emotions— figuring out what I thought they were feeling and then naming it.

"I do this all during the day," I said. "Sometimes with strangers in the street, or sometimes a cashier in a delicatessen. I focus on what they seem to be feeling. And then I name the feeling."

"Silently?"

"Right. Silently. *Naming* it seems to be important. Things change in my attitude toward them. I have more tolerance for boring people, or people who are bureaucratic. I try to identify what they're feeling—not what they're *thinking,* because that's pretty obvious."

"Because they're telling you."

"Right. Like if they say, '*Sign the form!*' it's pretty clear what they're thinking. But what emotion are they going through? Identifying it doesn't make me more sympathetic, but it does give me a chance to respond more appropriately. It kind of heads off impatience. And here's something odd. When I do this in conversation, something must be happening on my face, because

I see the expression on *their* face change. There's more connection. It's a subjective feeling, of course; there's no way to know if this is actually happening, but it seems to make a difference."

Matt looked at me for a moment. "You made this up?"

"Yeah." Long pause. "Does it relate to anything you've come across?"

"That's what I'm thinking." Another pause. Then: "It's related to the work we do training clinical psychologists. One of the things we talk about a lot is the therapeutic alliance. The bond between the therapist and client. Carl Rogers's whole approach was based on empathy. Reflecting and trying to build a therapeutic bond. We identify what they're feeling, and then respond, saying it back to them. We always assume that the stuff that matters when you're building that alliance is the step where I say, 'That must really be painful for you,' or 'Wow, that sounds really exciting.' But actually, first I have to do that in my own head: I have to say, 'What am I going to call this?' So, being nonjudgmental and positive might all start with internally stating the person's affect to yourself. I like this. I never thought of that piece of it. It's a very clever exercise."

Clever. What a nice word. I started to get excited.

Would it be possible to see if naming people's emotions could actually increase your empathy? "How would you study something like that?" I asked.

He briefly outlined a method where he could give people an app for their smartphones that they could tap every time they read someone's emotion and named it. They could do this for a week, and he could use standardized tools for testing empathy before and after a run. He thought for a moment and said, "I'd love to do that study."

I mentally noted his emotion and labeled it enthusiastic. I labeled my own as overjoyed.

I was eager to see if this naming game really worked. I asked Matt how much funding he might have to raise for a study like this. He thought for a second and made a rough guess. I realized I could fund it myself and save him months of writing grants. Almost immediately, though, I knew that I would want to write about the results in this book, and I wondered if my funding the study would in some way compromise it, making it seem less independent—like a drug company that advertises only its positive results. I agonized over this for a couple of weeks and finally concluded that, as a scientist, Matt would be just as interested in proving it wrong as proving it right, and that I would report on his findings no matter what they were—even if the results were totally null and my "clever exercise" failed miserably.

In fact, *I'd be happy to report on a failure,* I thought. *I have a lot of respect for failed experiments. It's how scientists know what doesn't work. I wish more so-called failures in research got attention. It would save others from going down the same dead-end alleys. It could be helpful to report on this if it didn't pan out.*

I was starting to think of the possible failure of the study as a public service, and we hadn't even begun it yet. *In science, as in art,* I thought, *you only arrive at success after you run the gauntlet of failures. People should be aware of that. I'd be* glad *to write about the failure of this idea.*

Anyway, that's how it seemed at the time.

LENDING MY BRAIN TO SCIENCE

I pulled into the Stony Brook garage a half hour early. I stood in the shade of a tree and waited for Matt to meet me and lead me to the lab.

He and his research team had worked out rigorous protocols,

and I was here to see what it would be like for someone to go through their study. I was going to be a mock subject. Whatever my result turned out to be, it would be excluded from the study, because I was, after all, not a naïve subject.

I knew, for instance, that I was not being put into a control group. But just to make sure I'd be as ordinary a subject as possible, I had stopped labeling other people's emotions for the previous two weeks, so I'd have a fresh start.

After a couple of minutes, I saw Matt waving as he came down the path. He introduced me to Tammy Rosen, a doctoral student who had helped design the study. As we walked to the lab, I wanted to find out as much as they could tell me without spoiling the experience of being an ordinary subject.

"How many people have been through the study so far?" I asked Matt.

"So far, twenty. Out of forty-five. Then we have the autistic population after that."

"So, how's it going?"

"It's very interesting."

"Really?" That word *interesting* could be positive or negative.

"I guess I shouldn't tell you too much, but it's exciting."

Exciting. That was definitely a positive word. I didn't ask any more questions, but I felt a little excitement myself. I realized, of course, that he might be so rigorous he'd be excited even by negative results, as long as they were definitive. But I hung on to my positive interpretation of the word. Not very scientific of me, but I couldn't resist.

We got to the lab and Tammy took me into a small room, where I filled out papers outlining my medical history, and then she started me on the pretrial tests. *A is to B as C is to what?* or, *What's missing from this pattern?* I love tests like this. I love them too much. I get competitive and want to ace them. I wound up taking twice as long as it usually takes to finish the session.

Next, they fitted me with an EEG cap, which involves a certain amount of poking at your scalp with a blunt object to ensure good electrical contact, and eventually leaves you looking like a robot with wires sprouting from your head. Scientists had recorded my brain waves with EEG machines many times during the days of *Scientific American Frontiers,* but never to test for empathy. This would be a new experience. I was helped across the room—we were trying not to detach my wires—and seated in a soundproof booth. There, I listened to recordings of the same phrase spoken happily, angrily, sadly, and in a few other ways I couldn't distinguish. My task was to name the emotions. Later, I saw pictures of people's faces displaying the same range of emotions, which I had to identify. It was fun, but occasionally confusing. How *did* these people feel? I was thinking maybe I wasn't as empathic as I thought I was.

I was exhausted by now. And this was just the pretrial run. I'd have to go through the same thing again the following week to see if my empathy had improved after seven days of reading and naming other people's emotions.

Tammy programmed my smartphone so I could send back a report every time I had an interaction with someone that lasted more than five seconds. I would have to silently figure out what emotion they were feeling, and then choose from a number of possible emotions on the phone's screen, or type in a more specific one of my own.

As soon as I got home, I started reading my wife, Arlene's face while we talked. After a while, she got used to my whipping out the phone while she was midsentence. The next day at lunch, I practiced on one of our daughters and her family. They asked questions about the study, and were amused when I'd pick up the phone and tell them, "I'm giving you a 'curious' for that." I was immersed in it.

The night of the second day, I had one of those dreams that seem so real you remember every word spoken by the dream's characters. More than that, while I was still asleep I could reflect on my behavior, as if I were inside the mind of the person in the dream, who was, after all, *me* inside my own mind. When I told Arlene about the dream, she smiled and said, "You're totally caught up in reading emotions, aren't you?" She was right. I was monitoring people's emotions in my sleep. Even my own emotions.

In my dream, I was visiting a mathematician I didn't know very well, who joked that we held different political opinions and we were probably headed for a good argument. "No, no," I told him. "I value the opinions of people I disagree with. It's the yin and yang of understanding. Reality is made up of opposing ideas." I even tried to explain all this using an equation. After a long and nonsensical tirade, I could see that his face had fallen. I could hear what the *me* in the dream was thinking: *You didn't show him how open you were. You lectured him.* All this attention to other people's emotions wasn't having much of an effect in my dream. I began to suspect that my unconscious was warning me not to get too excited. Things might not turn out so well with this study.

A WEEK LATER

I arrived at the lab early again, eager to see how high on the empathy scale I had gone after a week of labeling other people's emotions. I was fitted with the EEG cap and took the empathy tests again. Then I sat down with Matt and looked at my results. A small stone started forming in my stomach when I saw that I had scored as slightly *less* empathic than the week before. Matt didn't seem fazed by this. He was happy to see results, no matter what they were; whereas I was beginning to realize I was probably not going to win the Nobel Prize for my clever idea. He reminded me that the study was still going on and told me he was being very careful about not committing the error of peeking at the results before all the data were in. So, he hadn't begun to tabulate the results of the people who were truly being studied. I was, of course, just an observer. Still, Stockholm was drifting farther away.

Matt mentioned that the harder it is for someone to read their own emotions, the harder it is for them to read the emotions of others. Maybe I should have been working on becoming better aware of my *own* emotions.

I left the lab hoping Matt would have better results with the actual volunteers in the study. There still was hope because just a few days earlier I had read something Daniel Goleman, the pioneer in emotional intelligence, said: "The more sharply attentive we are, the more keenly we will sense another person's inner state." The volunteers would certainly be put through a week of being more sharply attentive. So maybe they would be a little more empathic.

And if not, I took comfort in the idea that I would be able to write an account of an experiment that took an interesting hypothesis and proved it wrong. This would be a helpful thing to do. It would be a service to mankind. In other words, I was slightly depressed.

NAMING EMOTIONS: DOES IT ACTUALLY WORK?

I began to wonder: *What if I'm kidding myself?* Then I got an email from Matt. He had some preliminary results.

I got to his office building early on a cold November day and found the door locked. I stood out in the icy wind for twenty minutes until I realized I was at the wrong building. I called him and in a few minutes he came smiling around the corner to fetch me. As we walked to the office, I was hoping I'd understand his results better than I understood his address.

We sat down in his office and I couldn't hide my wish for good news.

"To be clear," he said, "these are just the preliminary results. We still have more folks coming in. I'm not willing to etch them in stone, so to speak."

"Right."

"We have enough of the sample now that we're able to at least see some of the patterns that are emerging. A few of them really did surprise me."

"Like what?"

"One thing that was nice was that most people actually did the task," Matt said. "The range is really quite broad. One person reported in only twice during the week, but somebody else did it 132 times. . . . The average was 48. I would have been happy if we got once a day."

He reviewed the study with me. The participants were divided into three groups. The first group was asked to mentally figure out the emotion of the person they were talking to and use the smartphone app to name the emotion and send it in, the second group would just note the hair color of the other person and then send that in, and the third group would simply report that they had interacted with someone.

"I'm thinking about it as three different conditions that are decreasing in intensity in terms of what we're asking them to do. The improvement we saw was greater for the first two than for the third. Remember, the first two are really the more active ones. In the first two, you're saying, 'I want you to *notice* the person you're interacting with.' In both of those conditions, we saw greater improvement than the third condition, which was just noting there was an interaction."

So, the more *active* they were in studying the other person, the better they did on the empathy test at the end of the week.

And that brought him to another finding. "This is the one that really blew me away and surprised me," Matt said. "Let me show you this graph." He took out a piece of graph paper. "There's this wide range in terms of how much people actually logged in—how much they did this thing at all over the course of the week. The question is: If people do more of it, do they do better than the people who did the same thing but did *less* of it?"

"For instance," I said, "somebody who paid attention to someone's emotions 132 times . . . would they be expected to do better in the empathy test than somebody who did it only twice?"

"Exactly," Matt said. "And the answer is, they did."

The more they practiced it, the better they got at it: a "dose response relationship"—but *only in the condition where they paid attention to emotions and faces.*

THE BENEFITS OF PAYING ATTENTION

And then Matt noticed something else in the data. At first, he had assumed that identifying emotions and noticing hair color were two very different states of mind. But what if he considered them as similar? After all, both groups were being asked to do something that the third group wasn't: simply to pay attention to the person they were talking to. To *notice* something about them.

"The question that arose," Matt said, "was, *is there some-*

thing special that comes simply from that orientation, simply attending to the other person?"

The answer was yes.

His graduate student, Cara Keifer, analyzed the results and found that by deliberately paying attention, participants *do* improve in *one aspect* of empathy—their subjective sense of their emotional connection to others, or, as Matt said, "how much you feel that you're affected by other people's feelings." At the end of the week, when participants answered questions on how affected they were by others' feelings, without being told how they rated at the beginning of the week, their scores were higher.

In addition, something interesting emerged from the study about the two-way nature of communication. When someone speaks to us, we need to be alert to that person. And this readiness for communication was speeded up among the participants when they practiced paying attention. When Cara combined the data from Groups 1 and 2 (where they noted both emotions and hair color), the participants were ready to attend to the other person more quickly. As Matt described it, "This early, obligatory, social perception process [being alert to the other person] seems to happen *faster* after one week of [the experiment]."

This all sounded pretty good to me, but I wanted to make sure I wasn't reading more into it than I should. I asked Matt to put it as plainly as he could.

"To try to spell it out," he said, "noting and attending to emotions *does* seem to benefit one's ability to accurately read others' emotions after a single week, so long as you practice it every day (like with a muscle). On the other hand, simply paying attention to other people *at all* seems to make you feel more emotionally connected to them, and even primes your ability to quickly attend to social interactions as soon as they come your way."

So this idea of actively paying attention to the people we're communicating with not only makes intuitive sense, the results from this small study suggest it's actually not such a bad idea to give it a try.

There are even some simple tests you can take on the Internet to see if it's working.

CHAPTER 13

Working Alone on Building Empathy

THE "READING THE MIND IN THE EYES" TEST

In order to determine a person's level of empathy, Simon Baron-Cohen, the British psychologist, developed a test called Reading the Mind in the Eyes. You're shown a series of pictures of people in the throes of one emotion or another. You have to choose, from a list of four emotions, which one the person in the picture is feeling. Simple enough. Except that all you see is the person's eyes. It's not so easy when the emotions are subtle.

Researchers at Emory University used this test to see if meditation would improve people's ability to read emotions in other people's faces. Participants took the Reading the Mind in the Eyes test, then half of them received meditation training and the other half simply attended lectures on general well-being. After the first group had studied meditation, both groups took the emotion-reading test again. Those who had studied meditation had scores that were 4.6 percent higher than before meditation training. That didn't sound like much to me, although those who had not studied meditation showed no improvement at all. And, interestingly, while participants were reading emotions in the eyes, fMRI (functional magnetic resonance imaging) scans revealed that the meditators were more likely to show activity in the regions of the brain thought to be linked to empathy. It was a small study, but it suggested that more studies would be useful. So I decided to do an even smaller study and try it on myself.

I realized that a study involving one participant is not actually a study; it's more like a mental aberration, like taking your temperature every hour. But that's my style and I'm stuck with it. Unfortunately, we don't have an fMRI machine at our house, although if they were cheaper, I'm sure I'd have bought one. Instead, I would check my results just using the Reading the Mind in the Eyes test.

I took the test online, which anyone can do, and which is a lot of fun. There were thirty-six pictures of eyes expressing different emotions, and I identified thirty-three correctly. I was actually disappointed by this result, because I didn't see how a week or two of meditation was going to raise my score. It was probably already as high as it was going to go.

MEDITATION

There was another slight hitch. I didn't actually know how to meditate. Most of the meditation I had read about describes the process in strong religious or mystical language, none of which I was prepared to accept or even understand. One description of meditation began with a list of benefits that would accrue to my chakras. I had only a vague notion of what chakras were, and the description didn't help: "The chakra centers are like energetic motors within the mental/emotional/physical energy field we usually identify as 'me.'" This wasn't for the *me* I identify with as me.

I've never seen my liver, but I have an easier time believing I have one than that I have a chakra. No offense to people who are in touch with their chakras; it's just not for me.

I was going to need the help of somebody who meditated and who could describe it in terms I could understand. I needed a purely secular version of meditation.

In talking about it with friends, I discovered I already knew two people who meditated secularly. One was the actress Marlo Thomas and the other was the Supreme Court justice Stephen Breyer. They were from two different worlds, but they were ideal people to learn from.

Not only is Marlo a skilled actress, an occupation that requires empathy, she's also one of the country's top fundraisers. She and her staff raise almost a billion dollars a year for St. Jude Children's Research Hospital. To raise that much money, she has to be an expert communicator, able to explain the scientific dimensions of the hospital's research and to understand and appeal to the hearts of her donors. And their hearts' desires. Like salesmanship, fundraising only works when you keep in mind the other person's needs and not your own.

Stephen Breyer speaks openly about the need for empathy in judicial decisions. As he said in one interview, "When you're a judge and you spend your whole day in front of a computer screen, it's important to be able to imagine what other people's lives might be like, lives that your decisions will affect." In conversation, he's clear in maintaining that the awareness of how decisions will affect people doesn't diminish a judge's attention to the law. Rather, I get the impression he feels that a dose of empathy enhances the law and moves it closer to the greater good. This is what he means, I think, when he says, "This empathy, this ability to envision the practical consequences on one's contemporaries of a law or a legal decision, seems to me a crucial quality in a judge."

So, if there were two people from whom I could learn something about meditation, I couldn't do better than Marlo and Stephen. I talked with them separately and in each case the lesson was surprisingly short and to the point: Pay attention to your breathing. That was pretty much it, with a little advice on returning to concentrating on the breathing when my thoughts wander. I decided to sit quietly every morning for twenty minutes and concentrate on breathing in and breathing out. I'd see if a few weeks of that increased my empathy when I took the Reading the Mind in the Eyes test again.

Then I saw something online that made me think about trying to name other people's emotions again.

In a talk by Helen Riess, the psychiatrist who trains doctors to be more empathic, she suggested that we could increase our empathy by mentally labeling the emotions of others—just what I had begun doing. "When you're with someone," she said, "try labeling . . . is Jack upset? Is Jane excited? . . . It'll change how you hear what they're saying." This was almost exactly what I had come up with myself, and she was saying it makes a difference. I decided to keep doing it.

And there were other exercises that sounded promising.

In another part of her TED talk, Helen explores the importance of gazing into another person's eyes and the need we all have to be seen, to *know* we've been seen. The gaze changes us.

That struck a chord with me, because I know how important it is to look into the eyes of an audience when I'm giving a talk. I don't just scan the audience; I catch the eye of individual people and hold their gaze for a few seconds. When I do that, something happens between us. I'm actually talking to someone, not just saying the words I've prepared, and as a result something changes in my tone of voice. It becomes more personal and direct. And I get reinforcement from the warmth I see in their faces.

OXYTOCIN AND BONDING

I was beginning to see in what I was reading that eye gaze has an effect on brain chemistry that may be important to communication. Hundreds of papers published in the last few years have dealt with a molecule called oxytocin and its positive effect on us, some of which is specifically tied to eye gaze.

Oxytocin influences how much we trust other people and bond with them. The oversimplified moniker for oxytocin is "the love hormone." The most startling research (at least, the most startling to me) showed that when dog owners gaze into the eyes of their pets, the oxytocin levels in both the owners *and the dogs* increase. Does the jump in the level of oxytocin increase your overall willingness to trust in general? I don't know; but of all the possible ways to bring about a close connection, there would be none much odder than gazing lovingly at your dog in order to communicate better with your spouse. Still, there's something about gazing into the eyes of another person—or a friendly dog—that seems to be tied to bonding.

I was coming across an interesting collection of suggestions for how to raise your empathy level by doing everyday things— some backed by research and others just by intuition.

The magazine *Psychology Today* has recommended this: "Watch TV with the volume down and practice your nonverbal interpretation by reading what each character is feeling and talking about. This is best done with subtle dramas, not action movies." I was already doing something a lot like this. I was watching crime shows from Scandinavia. Unlike our crime shows in the United States, series like *Wallander* and *The Bridge* from Sweden have characters with a rich emotional life. Even the villains have families they care about and sometimes cry over. Watching those shows, I was spending an hour a day studying the actors' faces, reading their emotions and generally trying to figure out what's going on with them internally: Who's lying? Who's truly suspicious, or just a Swedish red herring? I rated my success by how early in the show I could figure out who was up to no good.

I do know how strange this sounds. And possibly useless. But it's fun. What's harder for me is the suggestion a number of people have made that a good way to increase one's empathy is by reading fiction—not potboilers, but good, solid, literary fiction, where the author delves into the emotional life of the characters with depth and sensitivity. And studies have shown that reading literary fiction improves Theory of Mind, as well.

This sounds like it ought to work, but I'm impatient reading fiction. So much so that, in order to force myself to read it, I belong to a book club that only reads novels. Otherwise, I would probably stick with books about science and history. The other members of the club, who are all highly articulate and literary, can't understand this blind spot of mine.

"*Why* don't you like novels?" they ask me.

"Because," I say, "you can just tell they're making it up."

This is usually followed by a compassionate pause while they rearrange their silverware.

Certainly, there's fiction I love, and I ought to read more of it. Novels give us the chance to enter into another person's life, to feel what they feel, to see life through their eyes, to enter their inner world. It's probably for this reason that Justice Stephen Breyer loves Proust. He told the *New York Review of Books*,

"It's all there in Proust—all mankind! Not only all the different character types, but also every emotion, every imaginable situation. Proust is a universal author: he can touch anyone."

If this is true, and if to be touched is to be enabled to touch others, I ought to give in and try it. In fact, I decided to try all these suggestions at once.

I would contaminate my experiment in meditation by including all the variables: While I meditated daily, I would also silently name the emotions of others, gaze into the eyes of the passing pooch, read the emotions of actors in dramas, and spend an occasional wrenching couple of hours with Madame Bovary at her most distressed, along with Proust and his memory cookies.

After six months of all this, I took the Reading the Mind in the Eyes test again. The first time I took it, I scored 33 out of 36. This time I got 36 out of 36. This proves nothing at all, of course, and ranks on a scale of reliable evidence just above wishing. But I was delighted anyway. Anything that makes me even *feel* that I have more empathy at my disposal is okay with me.

Since I'm convinced that empathy is at the heart of communication, I of course want more empathy. But that's not because I think empathy will cure the ills of the world.

In fact, sometimes empathy worries me.

CHAPTER 14

Dark Empathy

At forty-nine, Bernard Hopkins was a boxer still trading punches in the ring and still winning fights. They say he did it partly by perpetually staying fit, but mostly, having studied Sun Tzu's *Art of War*, he did it not by out-punching other boxers, but by defeating their strategies. He could anticipate what they had in mind and then frustrate them. Keeping an eye on his opponent's front foot, looking for it to lift off the floor, he knew in advance what the other boxer was planning to do next.

In a profile in the *New York Times Magazine*, Carlo Rotella said of him, "Figuring out what the other guy wants to do and not letting him do it is a matter of policy for Hopkins." He won matches not so much by knocking heads as by reading minds. Like a good communicator, he could read the body language of an opponent and know what he was thinking. He was able to attack an opponent's strategy even before he had to attack his person. He did this by careful study.

His mentor, Sun Tzu, had the same idea. He wrote, "If you know the enemy and know yourself, you need not fear the result of a hundred battles. If you know yourself but not the enemy, for every victory gained you will also suffer a defeat. If you know neither the enemy nor yourself, you will succumb in every battle."

What Hopkins and Sun Tzu are doing here is using an awareness of another person's mind against him—not using it to sympathize with him, but to defeat him.

Empathy and Theory of Mind are not the same as sympathy. Tuning in to another person's thoughts and feelings is not necessarily a path to good behavior. There's a dark side to empathy. While knowing what's happening in other people's minds can lead to bonding and caring about them, it doesn't have to. It can be used to keep others submissive.

The stereotypical view of empathy is that it makes you soft, that you have to abandon it if you need to be tough. On the contrary, when you have to be tough—or even if you choose to be cruel—empathy can be a useful tool. It doesn't necessarily make you a nice guy.

Bullies seem to know instinctively how to hurt you, how to make you feel defenseless and weak. They know what you're feeling and they can play a sonata on your tender feelings as if you were a violin.

Even normally empathic people can be moved to deliver unusual punishment, and apparently it doesn't take much effort. In a classic experiment conducted by Al Bandura in 1975, college students were told they were going to take part in a group task with students from another school. As part of the experiment, they were also told they'd be delivering electric shocks to the other students. One of the groups being studied heard an assistant calling the other students "animals" and another group heard an assistant referring to them as "nice." That slight change in language led to students' delivering higher levels of electrical shock to those they had heard referred to "animals."

But we don't have to go back to 1975 to see the misuse of empathy.

Reports have come out of Guantanamo that psychologists advised jailers there on how to make their prisoners feel helpless and out of control during "interrogation" sessions. In many cases, they did more than advise.

According to a U.S. Senate report, two psychologists were paid $81 million between 2006 and 2009 to devise and take part in a program that used theories of "learned helplessness" during what were regarded by many as torture sessions. According to the report, one of the psychologists running the program, James Mitchell (also known by the pseudonym "Grayson Swigert"), "had reviewed research on 'learned helplessness,' in which individuals might become passive and depressed in response to adverse or uncontrollable events. He theorized that inducing such a state could encourage a detainee to cooperate and provide information."

In both overseeing and conducting the sessions, psychologists were using their understanding of the inner life of detainees to disable them emotionally. They weren't just reading minds; they were breaking and entering.

Bullies and interrogators, maybe even psychologist/interrogators, might be expected to misuse empathy, but even large, trusted organizations have used empathy in an unethical way.

Merck and Company, the American pharmaceutical company, has a distinguished record of innovative research that has saved millions of lives. But for a time, its record of doing good was marred, and it was marred in part by how they misused empathy training.

In 2005, a U.S. congressional committee issued a memo describing the sales practices at Merck and Company. The company had stopped selling the drug Vioxx a year earlier after it became widely known that the drug was associated with strokes and heart attacks. But according to the memo, millions of prescriptions were written even as evidence mounted that the drug was unsafe. In fact, the company, with three thousand trained salespeople, "prohibited the representatives from discussing contrary studies (including those financed by Merck and Company) that showed increased risks from Vioxx." What strikes me most about this report is the detailed account of the sales training that intentionally used techniques of empathy and trust-building to convince doctors to use the product.

Body language was addressed in minute detail. According to the congressional memo, "Merck [and Company] representatives were taught how long to shake physicians' hands (three seconds), how to eat their bread when dining with physicians ('one small bite-size piece at a time'), and how to use 'verbal and nonverbal' cues when addressing a physician to 'subconsciously raise his/her level of trust.'"

In a course called "Captivating the Customer," representatives were asked to become familiar with "nonverbal techniques involving the eyes, head, fingers and hands, legs, overall posture, facial expression, and mirroring."

Mirroring. The very thing we teach as a basis of good communication, and which, used correctly, can help physicians care for their patients.

In notes for the leaders of the course, the concept of mirroring was explained this way: "Mirroring is the matching of patterns; verbal and nonverbal, with the intention of helping you enter the customer's world. It's positioning yourself to match the person talking. It subconsciously raises his/her level of trust by building a bridge of similarity." This is not bad advice for selling something, but not so good when you have evidence that what you're selling may be harmful.

Merck and Company paid a fine of $950 million. We can hope this is all behind them now and that the company has returned to its former laudable self. But, as the congressional committee showed, even trusted companies can slip, and their distinguished record can be crossed for a time by the shadow of Dark Empathy.

So, I don't think of empathy as a cure-all. It's a tool that can be used for good and for bad. Houses have been built with hammers and people have been murdered with them. Once radioactivity entered our lives over a hundred years ago, we had a tool that could diagnose and treat cancers, but, in time, could also annihilate whole cities.

Even when we think of empathy as a tool for good, it might not be a good idea to oversell its strengths, and we should remember that there will always be people who will use it against others for their own benefit.

The psychologist Paul Bloom takes a pretty dim view of empathy, mainly because he doesn't see it as leading inexorably to good behavior. He's pointed out that as far back as Adam Smith, writing in 1859, we've been aware that, as Smith said, "persons of delicate fibres" seeing a beggar's sores "are apt to feel an itching or uneasy sensation in the correspondent part of their own bodies." In a way, he was describing neural mirroring.

But Adam Smith and Paul Bloom both take the position that even though we might have a similar sensation, we won't necessarily act on our awareness of a *stranger's* misfortune. We tend, instead, to let our empathic powers drive us to action mainly in favor of people we know and to whom we're related. Bloom is pessimistic that increased empathy will lead to altruistic public policies or a good moral life, partly because identifying a single victim of a tragedy arouses us more than faceless multitudes do. We'll be concerned by the plight of the girl trapped at the bottom of a well, but give far less thought to millions of children dying of hunger or caught up in genocide. For Bloom, empathy is harder to achieve than we realize, and it doesn't lead to moral behavior or good policy as much as rationality does.

This may be true. But let's not forget there's a baby in the bathwater. Let's not dump empathy because it can't fix everything that's wrong with everybody. Empathy can be a useful tool for communication without saddling it with the responsibility of being the golden road to the good life. In any case, empathy is not going away. It's part of who we are.

As Bloom says, "Our hearts will always go out to the baby in the well; it's a measure of our humanity. But empathy will have to yield to reason if humanity is to have a future."

Maybe so.

It's true that the baby-in-the-well phenomenon might divert us from rational policies that affect faceless millions. But it's also the kind of focused, personal image that can help in communicating the very *need* for those policies.

A single human story can give flesh to numbers that are numbing.

Charities will put a personal face on hunger to bring attention to starving children in other countries. Instead of appealing to a rational interest in anonymous multitudes, they send us pictures of specific children, and if we contribute, sometimes it's even arranged for us to get thank-you notes from the children. The suspicion that some of these thank-you notes might be churned out on an assembly line doesn't seem to stem the flow of our empathy.

These appeals are directed to us as individuals, but it can go further. Specifying sorrow can capture the attention of whole nations.

In the summer of 2015, hundreds of thousands of people were fleeing from war. One family, having escaped from Syria, paid smugglers for a motorboat to take them from Turkey to Greece. But what the smugglers gave them instead was a rubber raft. With no other choice, they tried to use it. High waves spilled the family into the ocean, and the father tried desperately to hold his children's heads above water long enough to keep them from drowning. But when he was washed ashore, he was the only member of his family alive. The body of his three-year-old son, Aylan, was lying facedown on the sand, the edge of the ocean that had killed him lapping gently on his cheek. A photographer shot a picture of Aylan and within hours it rocketed around the world on the Internet. Immigrants, suddenly, were no longer statistics. Aylan was their face.

Canada, where Aylan's family had relatives ready to vouch for them and take them in, had refused them entry because of a missing paper establishing their immigrant status—a paper almost impossible to obtain in Syria. Now the phone lines and in-boxes of the Canadian authorities were flooded with demands for a more humane policy. In France, the day the picture appeared, President François Hollande announced that he had joined German chancellor Angela Merkel in proposing the European Union take in more refugees and distribute them among the twenty-eight member states. "Europe is a set of principles and values," he said. "It is time to act." He proposed taking in fourteen thousand migrants. In Britain, Prime Minister David Cameron, who had previously rejected Merkel's call for a system of immigrant quotas among European countries, changed his mind the day after the picture hit the Internet. "As a father," he said, "I felt deeply moved by the sight of that young boy on a beach in Turkey. Britain is a moral nation and we will fulfill our moral responsibilities." Britain was, of course, a moral nation before the picture appeared, but it took emotion and empathy to ignite that moral stance and turn it into thoughts of action. He proposed taking in twenty thousand people.

Pope Francis made the same distinction between numbers and personal imagery when he addressed Congress a month or so later, speaking about the refugees flooding Europe: "We must not be taken aback by their numbers, but rather view them as persons, seeing their faces and listening to their stories."

The promises to accept thousands of immigrants, made in emotional moments, weren't always kept at the promised level. As weeks passed and the emotion died down, there was plenty of slippage. But the emotional response to that individual story did play a part in policy decisions that might not have been made otherwise.

Bloom acknowledges that it's not merely a rational aware-ness of others' needs that moves us to good behavior. As he says, "Some spark of fellow feeling is needed to convert intelligence into action." I think he's right about that.

There are times we know what the rational action should be, but don't take it until we consider what the other person is feel-ing. I know, in my own life, I sometimes respond to a question with an answer that isn't really helpful. "Have you seen the can opener?" is not fully answered by saying, "No, I haven't seen it." The other person is still at a loss. I know it seems obvious, but sometimes remembering what it feels like to be facing a can without an opener can produce a little spark of empathy. If I respond to that spark, I might add a few words: "Maybe it's in that other drawer with the soup spoons." Boom. I'm cooperat-ing, and the spurt of reward hormones in my brain is a sign it's been worth the effort.

But as good as those reward hormones feel, I'm not thinking, in this book, of empathy as the basis of good behavior or morality; I'm looking at it as a tool for communication. I think it's an essential tool, and while it can be misused, it can help us make those important connections that lead to understanding.

The baby-in-the-well reaction is exactly the feeling we want the doctor to have for her patient (as long as she isn't swamped by it) so that her patient knows she sees and hears her. It's the way every communicator should relate to his audience: with attentiveness, focus, and total listening—sensing what the other person might be going through as he tells his story.

Relating is everything.

CHAPTER 15

Reading the Mind of the Reader

I know it sounds odd, but we've found that it's possible to have an inkling of what's going on in the mind of our audience even when they're not actually in the room with us—like when we write.

IMPROV AND WRITING

Along with improvisation, we teach science students how to distill their message: to get to the point right away, to not get lost in the details, to keep it clear and vivid, to make us care.

But we noticed that if we taught our students improvisation *before* we worked on writing, the writing sessions went more smoothly. The students were better able to think about whether or not the reader was following what they had to say, because they had been trained to think about the other person's state of mind—not just what they wanted to tell them.

So we took a cue from that and tried incorporating improv

games into the writing classes themselves—and the classes went even better.

The more we reinforced our students' ability to focus on the other person, the better able they were to express themselves with words that would land on the reader with clarity. Improv was, in a way, preparing them to read a person who wasn't anywhere near them in time and space.

But without being able to observe the reader's body language or tone of voice, a writer doesn't seem to have much to go on—except for what George Gopen believes all readers have: expectations.

GEORGE GOPEN: EXPECTATIONS

Gopen is Professor Emeritus of Rhetoric at Duke University. He says that when people read, they have some basic expectations that have to be met, or the reader will become confused and frustrated.

If he's right, then being aware of when we violate those expectations could give us a window into the mind of the reader almost as effectively as if we were observing them. We could pay some attention to the other person, instead of focusing so much on what we want to pour into their head.

After I read a paper George Gopen wrote with Judith Swan, called "The Science of Scientific Writing," I wrote Gopen a fan letter. I told him I'd be coming to Duke University soon to shoot a story for a science show and asked if we could find a few minutes to talk about his work.

He offered to pick me up at the Raleigh-Durham airport and, thanks to some terrible traffic, we spent two hours talking in his car. The conversation was stimulating. As we drove through the knotted traffic, he untangled the common sentence for me.

His idea is that the reader expects thoughts to be laid out in a certain order, and that affects how the reader reacts.

The Sentence: Expectations

The very beginning of the sentence, Gopen says, is where the reader expects to hear what the sentence is going to be about. If the writer doesn't get around to that until somewhere in the middle of the sentence, the reader will have to go back to the beginning, trying to figure out what the writer is talking about. As Gopen says, "Readers expect [a sentence] to be a story about whoever shows up first."

I'm an actor, so for me Gopen is describing a sentence as if it were a stage play: When the curtain goes up, the actor playing the leading part had better make an entrance pretty soon, or the play won't seem to be about Hamlet—it's liable to look like a play about Rosencrantz and Guildenstern.

Once the leading character is onstage, he'd better start *doing* something, or we'll wonder why he's there, and Gopen feels the verb—the action—should come soon after the hero's entrance.

The Stress Position

According to Gopen, readers assume that what comes at the end of the sentence has special importance. He calls it the *stress position,* a place of emphasis.

For me, the end of a sentence has the same place of honor as the punch line at the end of a joke.

It's kind of obvious, but suppose you were to tell a friend this joke (it's kind of a meta-joke):

A priest, a rabbi, and a minister walk into a bar, and the bartender says, "What is this, a joke?"

If you find that funny enough to tell your friend, it's a sure thing you wouldn't tell it this way:

A bartender says, "What is this, a joke?" because a priest, a rabbi, and a minister have just walked into his bar.

The setup comes first and the funny part—the twist—always comes at the end.

If the reader actually has these and a host of other expectations, then being aware of them might give us a clue about what the reader will be going through. Including distracting thoughts, like, *What's this guy driving at? Why am I suddenly not interested in this? Did I leave the stove on?*

But are we actually *able* to get into the reader's mind?

Not everyone thinks so.

THE READER'S PERSPECTIVE

In his elegant book *The Sense of Style,* Steven Pinker says that to write as if the reader were looking over your shoulder is probably not possible. It's just too difficult to take on the perspective of another person.

I wonder.

I think novelists do this when they draw convincing characters. I think an actor does it every time he or she walks onto a stage as a character who is clearly not him or her. There wouldn't be any theater if actors couldn't in some way take on the perspective of another person.

If the director calls "Action!" and I walk into a room to hang somebody up by a chain and torture him, I don't just walk in knowing I'm going to do this awful thing; I have to see it through the eyes of the character. I have to know *why* I'm doing it. More than that, I have to know that I *deserve* to do it. The character's reasoning becomes my reasoning. I'm not telling you *about* the character; for a few minutes, it's as though I *am* him.

It takes training and practice, of course, or everybody would be an actor. But the point is that it's possible, to some extent, to take on another person's perspective.

EVEN REACHING US WITH MATH

My friend Steve Strogatz works this way writing about mathematics.

Steve wants to convey not just math but the beauty of math. And he does it not so much by thinking about math, but by thinking about thinking—what the *other* person is thinking and feeling. In his essay "Writing about Math for the Perplexed and the Traumatized," he says:

> Explaining math well requires empathy. The explainer needs to recognize that there's another person on the receiving end of the explanation. But in our culture of mathematics, an all-too-common approach is to state the assumptions, state the theorems, prove the theorems, and stop. Any questions?
>
> What makes this approach so ineffective is that it answers questions the student hasn't thought to ask.

Instead, Steve engages the reader as a friend. He wrote a whole book about math as if he were actually writing to one particular good friend of his. I know this is true, because I'm the friend. He knows what I'm probably thinking when he writes a sentence, because he's spent hours with me trying to get me accustomed to basic ideas in math, like irrational numbers, or the fact that there's more than one kind of infinity. (How many kinds of infinity do we need, for God's sake? Isn't one that goes on forever long enough? Apparently not.)

But whether he's writing with me in mind or a stranger, Steve thinks about what the reader is probably thinking.

You may wonder why I'm going on about this. If you don't do a lot of writing for a living, you may be asking yourself how all this applies to you.

I think it applies to everyone, because we all write now. We're emailing, blogging, texting. And many of us still use whole sentences. We're applying for jobs in business letters, or applying for love on dating sites.

My guess is that even in writing, respecting the other person's experience gives us our best shot at being clear and vivid, and our best shot, if not at being loved, at least at being understood.

CHAPTER 16

Teaching and
the Flame Challenge

MAKING IT CLEAR TO ELEVEN-YEAR-OLDS

In teaching, everyone seems to have heard the phrase "start with what they know." But it's another thing to know how to do it. I had a teacher once who seemed entirely innocent of the concept.

When I was eleven years old, I was beginning to try to figure out how things got the way they were. I was particularly amazed by the flame at the end of the candle. Why did it give off light like that? Why was it hot? And how come it wasn't solid? If you moved fast enough, you could put your finger right through it.

That year, I had a teacher I liked very much. She was cheerful, young, and had an opulent chest. These were qualities that at the time recommended her highly to my curiosity. So one day I asked her, "What's a flame? What's going on in there?"

She paused for a long moment, searching her mind for an accurate answer. Finally, she said one word: "Oxidation."

That was it. Oxidation. I hadn't known what a flame was, and now I didn't know what oxidation was either. I still had a crush on her, but I was back where I started. Except now I had another name for it.

Simply giving a name to something doesn't explain it, as much as we might like it to. Richard Feynman told the story of a boy who taunted him when he was young, challenging him to name a bird they'd just seen. When he couldn't, the boy said, "It's a brown-throated thrush. Your father doesn't teach you anything!" But his father already *had* taught him something that was more important than the name of a thing. He'd said to him, "See that bird? It's a Spencer's warbler." Feynman could tell his father had made up the name. Then he started to give Richard fictitious names of the same bird in several languages. "You can know the name of that bird in all the languages of the world," he said, "but when you're finished, you'll know absolutely nothing whatever about the bird. . . . so let's look at the bird and see what it's doing—that's what counts."

Many decades after my teacher had renamed a flame for me, I was writing an article about communication for the journal *Science,* hoping to stir scientists to reach out to the public with the story of their work. I was about halfway through when I realized I was writing a litany of familiar arguments. It wasn't personal. It didn't have a story; it was just the facts, which by themselves can be soporific. I wasn't following my own advice. *There must be some personal angle I can come in on,* I thought. And then, suddenly, I remembered: *oxidation.*

I went back to the beginning of the piece and told the flame story. My curiosity, my teacher's frustrated attempt to explain fire. By the time I got to the end of the article, I realized I had a way to engage the scientists I was writing for. I could invite them to join a contest. I asked, "Would you be willing to have a go at writing your own explanation of what a flame is—one that an eleven-year-old would find intelligible, maybe even fun?" And then came the best part: The contest would be judged by real eleven-year-olds.

I don't know who came up with the idea of eleven-year-olds as judges. Did I think of it, or was it Howie Schneider, or Liz Bass, the director of the Center for Communicating Science at the time? It seemed like a fun idea, a cherry on the sundae, but it turned out to give insights into teaching. The kids were very much in charge of their own learning as they judged the entries, and that seemed to make a difference.

They loved the idea that they could say to an adult, even an expert adult, "That's a good explanation, but it could be a little better." And in order to know whether an entry *could* be better, they had to become familiar with what was actually going on in a flame. Their teachers could work with them on that, and they were motivated to learn because that gave them the background for judging. Then, what they learned from their teacher was reinforced by reading several entries and comparing them. They were becoming specialists in what goes on in a flame, and it was fun, even though it was a complicated question—much more complicated than I knew at the time.

I hadn't realized when we started the contest that about 150 years earlier the great scientist Michael Faraday had delivered nine lectures to young people on the subject of what's going on in a flame. The lectures fill a book. And the question is even more complicated than that. Science had not yet thought up the puzzling intricacies of quantum physics. If they had, Faraday's book would be even fatter. And here I was, asking scientists to explain a flame to schoolkids in a few hundred words or a short video. But both the scientists and the kids were having serious fun. Some of the kids' deliberations were shot on video, and I could see in them a sense of purpose that matched any of the boards of directors I've sat on. One boy, criticizing a video he thought was a little hokey, said, "It's okay to be funny, but you don't have to be silly. We're eleven—we're not seven." Another example of the importance of truly knowing your audience.

The contest not only gave the kids authority over the scientists, it also meant they had to listen to one another's reasoning as they discussed how they should vote. The learning process was a group activity and they thrived on it. In one class, students told their teacher, "I wish we could learn everything this way."

We had kids from across the United States and a dozen other countries judging entries from around the world.

One of those entries in the first year of the contest came from a young American studying for his doctoral degree in Austria. He'd heard about the contest in a podcast of the radio show *Science Friday*. He immediately knew he wanted to enter the contest and decided the most engaging way to explain a flame would be to build a humorous video. But he had caught up with the podcast so late that he had only two weeks before the deadline to come up with his video. And he was thinking of an elaborate production that involved writing a new song and creating animations.

His work in the lab involved repairing a broken piece of equipment, which didn't seem nearly as interesting as putting together an entry for the Flame Challenge, so he told his boss he was taking two weeks off to work on his entry for the contest. He went home and explained to his wife and young daughter that they wouldn't be seeing much of him for the next couple of weeks, and then he locked himself in the basement.

During the next two feverish weeks, he wrote a song, arranged it and performed it himself, wrote a script and performed it, and drew elaborate animations that compared the atoms in a flame to the unusual image of Legos punching it out in a boxing ring.

At the last moment, only minutes before the deadline, he posted the final version of his video. But the next day, he realized that the entry hadn't gone out and he had missed the deadline. He sent a heartfelt appeal to the Center, explaining that a technical glitch had caused the delay and begging for the chance to have his entry considered.

It *was* considered, the kids judged, and Ben Ames became the first winner of the Flame Challenge. His song became a hit in dozens of classrooms. Surprisingly, the lyric included some fancy technical words, like *pyrolysis, incandescence, chemiluminescence,* and my old favorite, *oxidation.* But amazingly, the kids weren't put off by these words; instead, they loved them. One of the students said later that even when she wasn't thinking about a flame, the song would be running through her head and she'd be learning from it.

We brought Ben to New York to announce his win at the World Science Festival—and, as a surprise, a classroom of kids who were his fans (and judges) came on stage and serenaded him with what was now their favorite song about pyrolysis.

Online, Ben's video brought more attention to the Flame Challenge, and to Ben himself. He was asked to help create a pilot for a television series of animated science shows for kids. The contest was spreading the notion of science communication in more ways than we had imagined it would.

As the contest grew, the American Chemical Society and the American Association for the Advancement of Science became sponsors. It had begun as an improvised attempt to humanize an article and was taking us to unexpected places. Like a true improvisation.

AUTONOMY

In the second year of the Flame Challenge, we took a step further toward putting the kids in charge. We began asking *them* to come up with questions for scientists. In the first six years of the Flame Challenge, the questions they pointed us to were, "What is a flame?" "What is time?" "What is color?" "What is sleep?" "What is sound?" and "What is energy?" What began as an exercise for scientists to explain something complex with clarity had surprised us by becoming a learning experience both for the kids and for those of us who had started the contest.

If the first principle of teaching is to start with what they know, I think the Flame Challenge suggests that next in importance is that a little autonomy can give students the joy of discovery. And both of these ideas involve empathy and Theory of Mind—a recognition of the interior life of the student: being aware of what they know—and what they *want* to know.

My friend Steve Strogatz had been discovering much the same thing while teaching math. Steve is an award-winning teacher, and yet he had become increasingly dissatisfied with his usual way of teaching, which was lecturing. I wasn't sure why he was dissatisfied. I've watched Steve lecture and it was delightful to see how clear and engaging he was. Still, he felt his students were enjoying the lecture as a performance, then doing their homework, but not really becoming engaged deeply with the math. He wanted to find a way to get them more involved and began experimenting with a kind of "active teaching" based on the students' own sense of inquiry.

He tried it out in a course he refers to as "Math for People Who Hate Math." "These were seniors," Steve said, "who had put off fulfilling their 'math and quantitative reasoning' requirement for as long as they possibly could. The only thing standing between them and graduation was this course and the swim test."

Instead of lecturing them, and instead of showing them how to solve problems and come up with The Answer, he gave them intriguing puzzles and let them come up with their own solutions.

For instance, you're in his class, your first math class in college, and he gives you a piece of paper and a pair of scissors and says, "Fold the paper so that you can make just one cut across the paper with the scissors and you wind up with a piece of paper shaped like a triangle."

Not so easy. Is it even possible?

It is, but you and your classmates have to figure it out on your own. One person will get an inspiration for an approach to the problem, and even if it doesn't work, another will take the idea and build on it.

Yes And, with scissors.

There's a sense of collaboration and fun. The class may not realize it at first, but they're doing math and quantitative reasoning. They do it happily because, like the kids in the Flame Challenge, to a great extent they're in charge of the process.

There's pleasure for the teacher, too. "Teaching this class," Steve says, "was a joyful experience surpassing any other in my career. . . . I certainly love teaching students who already like math. But there's something remarkable about working with a group of students who think they hate math or find it boring, and then turning them around, even just a little bit."

This sounded to me like the reward that comes from paying attention and noticing what the other person is going through: that little spurt of happiness hormone.

For Steve, it's a necessary part of teaching. He said, "I need to know what works for you. And I won't know unless I'm empathic. There's no other way. What else can it be, if it's a mind communicating with another mind?"

GETTING PERSONAL IS A TWO-WAY STREET

Sometimes, being willing to see the other person means you have to be willing to let them see you. I saw an interesting example of this when we began training teaching assistants. TAs, as they're called, are graduate students who are assigned the task of teaching a course to undergraduates. Typically, they know their material extremely well, but they have little or no experience in communicating what they know. This can defeat one of the reasons for having them teach undergraduates in the first place: the hope that they'll give their students such a fascinating view of biology or physics that undergraduates will be inspired to study those subjects themselves. Too often, though, the undergraduates are scared away by boredom and the frustrating sense that they just don't get it and never will.

We started a pilot program, training a group of TAs teaching biology. Rounding them up was tough going. Graduate students have a heavy academic load, and when they have to take the time to teach classes, they're not in the mood to take more classes on how to do it. We limited the training to one hour a week, hoping that would give us enough time to be of help.

One of the main things we wanted to do was to get the TAs comfortable with actually making contact with their students. They had to figure out if the students comprehended what they were being taught, if they understood the experiments they were doing, and if they were tuning in to the meaning of the facts and techniques they were exposed to. The way to do that, we felt, was to establish a personal connection with the people in the class. So the TAs were asked to make *at least one personal con-*

tact with their students before the next session. It could be utterly simple and pro forma, like "Hi, how are you doing?"

They weren't comfortable with this. More than one teaching assistant said it would be totally inappropriate to make any kind of personal contact with the students. They wanted nothing to do with it. Yet, how could they read their students' minds if they weren't going to pay attention to them? And how would the students learn if they were shut out as a mere anonymous audience?

Finally, following the lead of one of the more comfortable TAs, most of those in our class started to make contact and reach for a more personal connection with their students. For some of them, the results were memorable.

One TA wrote a report after a class that had been a turning point for him:

> I totally switched the mood in class yesterday. . . . As we switched gears to talk about their upcoming experiment, I told a very personal story that resonates so much with me. It was a story of an experiment I did with a beloved when my curious little scientist mind was blooming with so many creative ideas. I obtained two stethoscopes from my dad, who's a medical doctor. Then I placed the earbuds of one of the stethoscopes in my ears and the diaphragm against her heart so that I was listening to her heartbeat. She used the second stethoscope in a similar setup on me. The goal here was to see if listening to someone else's heartbeat would synchronize one's heartbeat with the other person's. We strapped a cardio-microphone to ourselves to measure the changes in heartbeat, and we closed our eyes as I held her left hand—for no particular reason. During that moment, it felt like I was feeling her being.

Every heartbeat struck like the footstep of life itself. She had never been more alive to me. When we analyzed the data, we realized that our heartbeats started off as disparate rhythms but synchronized over time. But afterwards, we discussed the fact that we shouldn't have held hands, since it introduced another variable.

We decided to repeat the experiment. Except that we never did. And we will never repeat it, because she died in a car accident afterwards.

I explained to the students that whenever I think about experiments and variables, this incident revives in my mind. I leveraged this story to talk about the importance of controlling variables and, of course, included a little life lesson I have learned about what it means to be alive and to be actively present in the present. The response from the class was an overwhelming somberness that zapped across the room.

After the class, a student came to me and told me she really appreciated how much I cared about them and that she feels bad when her work is subpar because she sees the energy I put into helping them to learn. That seemingly little comment was quite helpful in evaluating the impact I'm having.

MY FAVORITE IMAGINARY PICTURE

We keep looking for ways to give people in our workshops an experience that helps them crash the resistance so many of us have to being personal and vulnerable with others.

Val Lantz-Gefroh, the director of improvisation at the Center, invented an exercise for science students that she always ends her improv class with, and it almost always affects people emotionally. It's one where, like the TA, she reveals something personal, in her case by holding up a blank piece of paper and describing an imaginary family photograph. The page is blank, but as she describes the picture in detail and tells the story behind it, the people in the room begin to see the people in the image.

She tells the story of the day the photo was taken, and what the picture means to her emotionally. And then she goes a step further. She invites the class to describe their own imaginary pictures. And, one by one, they pick up the paper, show their picture, and open up in a way they never have before.

One student who until then had not been especially personal when talking to a group held up the blank paper and began describing a picture of his grandfather. It was an ordinary picture of a man sitting in a chair, looking at the camera, but his description, and his story, began to move him. Before long he was choked up and couldn't finish the description. It was a brand-new experience for the student, and it felt strange to him. After the class, Val said, "He just kept looking at me. He had discovered something in himself and he couldn't quite articulate it. It was as though he was cracked open as a person." Later, he wrote Val an email:

I loved the improvisation parts of the workshop, especially the picture exercise.

When I was a teenager I grew fast and was tall and rangy and felt out of control of my body a lot of the time. I loved swimming—especially that part after you dive in when you are flying under the water and you can turn over and look at your reflection in the surface, or curl into a ball, or flip over. You are not constrained by the usual two dimensions of land.

The improv exercises were the emotional and intellectual equivalent of that. They made you free to get out of your own way and connect to your self.

And by connecting to your self you can connect to your audience.

On a personal note, I left the workshop, drove in the rain to the top of the biggest local hill, and stood on the roof of my car watching the passing storm. I don't usually do that.

I'm not suggesting that scientists get so moved by their own inner life that they get choked up in public, but in a controlled workshop setting, experiencing a moment of emotion in front of the class can be liberating—for the speaker and for the other people in the room as well. I walked into a workshop session once when they were halfway through the picture exercise and sat down next to a senior scientist. I asked her how it was going. She turned to me with moist eyes and said, "Great. We're all in tears!"

I don't have proof of this, but I think I've seen that when we're able to open up like this with people we've just met, it can be easier later to leap into a relationship with an *audience* we've just met. Or with a class that we see every day. We can reach more freely for emotional images, even about technical things.

I actually do think there are times when technical descriptions can be cast in emotional terms. It has to be crafted for the right audience, of course, and it has to be accurate, but once we've become accustomed to brushing shoulders with the emotional, it can make a difference.

For instance, here are two ways to describe the hugely important moment in science when the Higgs boson was discovered. Even a very few words with an emotional tinge can make a difference.

On the one hand, you could say it like this:

After years of work, scientists at CERN, in Switzerland, recently made a highly important discovery: the confirmation of a particle that had been proposed theoretically decades earlier. Here's what they found and why it was considered so important.

Or you could say it like this:

Scientists at CERN, in Switzerland, broke out the champagne and hugged one another in celebration. They had discovered something whose possible existence had tantalized them for decades. Here's what they found and why it got them so excited.

Each paragraph contains exactly the same number of words. But the second one, which is an entirely true and accurate account of what happened that day, includes just a few words that excite a little emotion in the reader: *broke out champagne, hugged, celebration, tantalized, excited.*

You might feel that even this much emotion cheapens the account. I would agree if this were a technical report in a scientific journal. But for us laypeople, a couple of emotional words can often turn a recital of the facts into something more engaging that will stick in the mind. And that's the interesting thing about emotion. It doesn't just make what we say more engaging—it makes it more memorable, too.

And why tell them something we think is important if we don't want them to remember it?

CHAPTER 17

Emotion Makes It Memorable

". . . A kiss is just a kiss," according to the chorus of the old ballad. But Herman Hupfeld, who wrote the words and music, knew better. In the verse that we seldom hear, he reminds us that it's simple, emotional things we go to for relief from the big, heavy things.

> *This day and age we're living in*
> *Gives cause for apprehension*
> *With speed and new invention*
> *And things like fourth dimension.*
> *Yet we get a trifle weary*
> *With Mr. Einstein's theory.*
> *So we must get down to earth at times*
> *Relax, relieve the tension*

It's the small, emotional things that stick in the head. A kiss is just a kiss, but as one of "the fundamental things of life," it can not only give us *relief* from Mr. Einstein's theory, it's the kind of thing, odd as it may seem, that can help us *remember* his theory.

I was talking one day with the memory researcher Jim McGaugh, who turned to me and said, "I bet if I asked you to remember your first kiss, you'd be able to." I certainly could, and did, as soon as he mentioned it. But it wasn't some romantic moment in the amber light of the setting sun that I remembered. It was an awkward twelve-year-olds' spin-the-bottle smooch that didn't do much for either one of us. It was actually a little embarrassing at the time.

And that illustrated Jim's point: We remember things that are tied to an emotion. Any emotion, even embarrassment.

"I'm sure," he said, "that all Nobel Prize winners know exactly where they were and what they were doing when they got the news that they won the Nobel Prize, and they'll never forget that. I'm equally sure that people who pick up body parts after an airplane crash will remember forever having seen that."

Before I met Jim McGaugh, I was aware that fear could lodge something in memory, like when we see a snake in the grass. But he's shown that events tend to be remembered when they're associated with any strong emotion: joy, shame, disgust—happy and unhappy emotions alike.

"It doesn't have to be frightening," he said. "It can be insulting. I can say to you—which is not true, but I could say—'You know, I read some of your stuff. It's not as good as it used to be. You might give that some thought.' Now, if you believed that I had said that, you'd remember that forever."

And I would. I can remember my first bad review as an actor, not just in general, but word for word. A bad review hangs on your brain by its pointy little claws, its stinger probing your amygdala for as long as you live. On the bright side, I also remember happy things, like almost every great dish of pasta I've ever eaten at home and abroad. There is, I'm certain, an evolutionary advantage in remembering a superb dish of rigatoni.

Jim McGaugh and I were walking along the tree-lined paths of his campus at the University of California, Irvine. We stepped into his lab, where his longtime collaborator Larry Cahill was setting up an experiment that we could film for *Scientific American Frontiers.*

A young woman named Malina was sitting in front of a monitor, looking at pictures that were emotionally arousing, like a gun, an injured boy, a decomposing dog, or a snake. The test was to see how many of these she would remember a week later. To help her remember, after she looked at the pictures she was asked to stick her arm in a tub of ice water. *How would that help her remember?* I wondered.

"The reason we're doing that," Larry said, "is this is a technique that's well known to activate the body's stress-hormone response. And we believe that the body's stress-hormone response acts to enhance memory." Still, that seemed to me like an extreme way to do it. Or, as Malina said when she dipped her elbow in the tub, "Oh! Oh!!!"

"How long can people take this?" I asked.

"Many people, myself included," he said, "are wimps that can take it for about a minute."

Malina took it for a full three minutes.

"Right now," Larry said, "her brain is busy storing all that information. We call it *consolidating.* Kind of like Jell-O setting. During the setting process, the stress hormones are working on the newly forming memory to enhance the storage of it."

After the three minutes were up, Malina took her arm out.

"How did the ice water feel?" I asked.

"Horrible. Absolutely horrible."

A week after the test, subjects like Malina who had the ice water treatment remembered the emotionally charged pictures better than those who hadn't plunged their arms in cold water. The stress hormones had helped solidify their memories.

This process of solidifying, or consolidating, their memories is necessary because without it, we'd all remember—with equal importance—everything we ever saw, heard, or dreamed. This would probably make it impossible to get through the day.

So, emotion helps us remember. That was clear. But this was new to me: that a bit of stress can help make that memory stick and feel more important than other memories.

This combination of emotion and stress may be why I can't forget a certain November day in Chicago. It's engraved on my memory.

I was getting ready to give a talk about communication at the annual meeting of the Association of American Medical Colleges. I had been looking forward to this. Hundreds of deans of medical schools would be there, and it would be a chance to tell them about the Center for Communicating Science. The stress, and the drama, started when acting commitments suddenly showed up. I'd been asked to do the play *Love Letters* on Broadway with Candice Bergen. I love the play and wanted very much to act with Candy Bergen, but now I was told that we were to open the same day that I was supposed to be in Chicago. I asked if we could open a day later. They said yes. Then the next complication arrived in my inbox. I had been playing a villain on the television series *The Blacklist,* and they were scheduling me for two shows to be shot while I was doing *Love Letters.* This wasn't going to be possible, so they asked me to please come in on the day I was flying to Chicago to shoot a scene where they would kill off my character.

So the schedule I had got myself into was to go to the studio in Brooklyn, have my head blown off by a pipe bomb, fly to Chicago, give a talk in front of twenty-eight hundred people, fly back to New York, and the next day open on Broadway. But the drama hadn't really started yet.

When I got to the hotel in Chicago, I asked for a wake-up call. The talk was early the next morning. I couldn't be late. "Right! Seven A.M.," the operator said. "We'll give you a call at seven sharp." At 8:00 A.M. I was awakened by a wave of panic. The phone hadn't rung. I threw on my clothes. No time to shower. And there was no deodorant in my suitcase. I have a pathological fear of smelling bad, especially in front of twenty-eight hundred people. I rushed downstairs and into the holding room next to the stage, where I noticed a cache of soft drinks and lemon slices. I put a lemon slice under each arm and went on stage. Filled with the spirit of improvisation, I gave the talk, and it went well. Five more medical schools affiliated with the Center, which thrilled me. I flew back to New York and twenty-four hours later walked out on stage with Candy Bergen.

This isn't a story that ranks with the discovery of penicillin or the moment that Einstein first thought of curved space, but for me, it couldn't have been more emotional, or more stressful, and I won't ever forget it.

Should we add a little stress when we tell stories about science, in order to lock them in our audience's memories?

I wondered what Larry Cahill would think about all that. It had been ten years since we filmed him demonstrating the ice water experiment. He might have come up with some new ideas since then.

I called Larry and asked if that extra bit of vicarious stress would help an audience store a memory better. He didn't think you had to go that far to prompt a memory. He said, "If you go from not very emotional to just slightly emotional in a story, you can see an enhancement of memory." So, the stress of putting your elbow in ice water helps form a memory, but Larry felt you don't have to put ice cubes in your story. You just need emotion.

But in the past ten years, Larry has found that there are differences in how effective some emotions are. "Emotion is not a broad-brush amplifier of all things," he told me. "It doesn't simply turn up the volume knob. The *kind* of emotion matters." Laughter, he's found, is far better than anger. And not just as an aid to memory, but as a way to connect. He's especially aware of this when he teaches.

Larry has twice won an award as the best teacher on his campus. I wondered if he had a deliberate strategy.

"Engagement. I engage them. I get them to care." And he has what he calls a secret weapon. "When I teach—it's almost an unfair advantage if you can get people laughing. It disarms them. It puts everybody in the same tribe. It gets their guard down for little while—and then you can slip something in."

"How do you know when you've got them engaged?"

"Well, you can see it in their eyes. They're focused. Paying attention, They're not checking their insta-twaddle accounts."

I think he's right about laughter, as long as it's unforced and comes naturally. (I have a painful set of memories of after-dinner speakers who were "reminded of a humorous story.")

But genuine humor and true, open laughter almost always lead to engagement. As Larry said, quoting the great Danish comedian Victor Borge, "Laughter is the shortest distance between two people."

LORNA ROLE

Someone who I assume doesn't have much trouble keeping her students engaged is Lorna Role. Lorna chairs the Department of Neurobiology and Behavior at Stony Brook University. I went to her office to talk about memory and emotion, and I found her to be a scientist with an unusual combination of deep knowledge and playful spontaneity.

When we talked about Jim McGaugh and his maxim, "Everybody remembers where they were when they had their first kiss," she remembered her own first kiss: "It wasn't very successful. We both had braces. . . . We locked."

"You *locked?*"

"It was not pleasant."

"So, I guess that's perfect. You'll never forget it."

"Never. It doesn't fit in the *nice* category, though. It goes in the *traumatic* category. But how amazing it is that a memory is more imprinted if the experience that surrounded the memory being formed was a very salient experience."

"So, *must* you have emotion associated with something to be able to remember it?" I asked. "Or is it just that emotion makes a memory stronger?"

"I think it's more a question of strength," she said. "And it's not a simple curve. More emotional context doesn't necessarily mean better or stronger memory."

And, apparently, some people find emotion in things that others don't.

"What about people who are trying to learn stuff that doesn't seem to most of us to have an emotional content?" I asked her. "What about you? Can you get emotional about something highly technical?"

"Well," she said, "if I was going to list where I was 'on the *day of . . . ,'* on my list would obviously be the day that Kennedy was shot—and 9/11—but one of them would be the first time I read the paper that described single-channel recording . . . a recording of a receptor opening and closing."

Later, after our talk, I would go home and look up what a receptor was (roughly speaking, a little thing on the end of a brain cell that receives what another brain cell is squirting at it, which is how our billions of brain cells talk to one another, making us feel hungry or sad, or Republican). I didn't want to interrupt her, though, so all I said at the time was, "You mean a receptor in the brain?"

"Yep. But it was the fact that you could *record* it. Neher and Sakmann got the Nobel Prize for it. I remember where I was standing. I remember what I was wearing when I opened *Nature* and saw those traces."

It was fun to see her excitement. This must be the kind of hair-rising-on-the-neck experience that scientists have when they know they're in the presence of something never known before. And the ability to describe the emotion they felt in that moment helps the rest of us understand why science is so exciting.

"What made you remember that so vividly?" I asked her.

"I got a rush of how it was a new world. A new era. It was a *molecule* opening and closing. It's the elemental unit of neural communication—the ultimate molecular biology. They had figured out a way to record just that electrical activity. Just that open and close . . . It was a very emotional moment."

"You felt it that strongly."

"That was 'Hallelujah Chorus' stuff. And I tell people if they don't hear the 'Hallelujah Chorus' when they do this work, they shouldn't be a biophysicist."

FEAR

As helpful as the thrill of discovery can be for forming a memory, fear still has a powerful effect on us. Lorna told me how she had worked on manipulating memories—looking for changes she could introduce into the chemistry of a brain, for instance, to retain a memory, or to get rid of one. And memories formed in fear were her original focus.

"That was the first thing I started with," she said, "because it's such a potent thing. It's so potent at imposing a memory that it only takes one trial: one pairing of a sound with a shock for the animal so that the next time they hear that sound they're like, *Oh, oh, shock.* It makes sense, right? You're going to learn if something's bad. It's not like you're going to say, 'Well, that lion was really unpleasant. I think maybe I'll come back here and see if I like it the next time.' You *get it,* right? It's not like you're going to get a lot of chances."

"And that's why people thought for so long that the one emotion that makes you remember something is fear?" I asked.

"Right, because it's one of the easiest ones to demonstrate. But both positive experiences and negative experiences can affect memory."

THE AMYGDALA

This activity was happening, she told me, in the little almond-shaped part of the brain called the amygdala, where both positive and negative experiences register. "Actually," she said, "they *both* happen in the amygdala, and people didn't appreciate that either; that the amygdala actually is really complicated and it involves not just fear learning but also positive-reward learning. It's a really interesting, very antique structure; it's really old. It's really, really fundamental."

I was feeling more than ever that emotion itself is an ancient, fundamental means of gaining and keeping knowledge. It may not be the *only* way we retain what's communicated to us, but it would be careless to ignore such a powerful tool when we want people to understand and remember what we tell them.

"So, how can we excite emotions in people who have no training in what we're talking about?" I asked her.

"Story," she said. "Like when you were sick—the anastomosis story."

She had participated in one of our workshops and heard me tell my story of nearly dying in a small town in Chile.

In an emergency operation in the middle of the night, a doctor had to cut out about a yard of my intestine to save my life. He was a brilliant surgeon and an extraordinary communicator. He realized he had to perform what's called an "end-to-end anastomosis," but he didn't use that term when he told me what he had to do. Instead, he leaned in, made eye contact, and said, "Something has gone wrong with your intestine, and we have to cut out the bad part and sew the two good ends together." I have never heard a clearer, more accurate description of something with such a frighteningly fancy name.

"Everybody remembers that story," Lorna said. "It's very empathetic."

The story is memorable because a life is in the balance. It will stick longer in the minds of the doctors in the audience than if I say, "When you talk to a patient, try not to use jargon."

If we're looking for a way to bring emotion to someone, a story is the perfect vehicle. We can't resist stories. We crave them.

CHAPTER 18

Story and the Brain

Don Hewitt, who invented the television show *60 Minutes,* had a story he loved to tell me. I knew him for thirty years and he probably told me this same story four or five times a year. I think it never stopped amazing him. It went like this: When a producer would come into his office to pitch a segment, if they started telling him about an issue, or a law that needed to be changed, or a scam that was making the rounds, he would put up his hand to stop them, and he'd say, "Tell me a story."

Don was certain that these four words were what kept *60 Minutes* at the top of the ratings for decades. He was probably right. The reports the show produced were usually serious and informative, but they didn't seem like lessons. A segment wasn't a treatise on espionage, it was a story about the fortunes of one particular spy; it wasn't an analysis of corporate malfeasance, it tracked the misfortunes of a whistle-blower. The show repeatedly won awards, and it won the allegiance of the audience.

When scientists are talking to the public, they're trying to reach that very same audience, and so are corporations and government agencies, and they probably won't reach them as well as they could without a good sense of story.

Our lives are filled with stories, and yet when we want to communicate something important to us, we often forget there's a story behind it. This is even stranger when you remember how deeply embedded storytelling is in all of us.

I got insight into this when I was talking about story one day with my friend Graham Chedd. Graham and I have been friends for more than twenty years, and worked together to help scientists communicate clearly on the science programs we shot for television. As producer and director, Graham had me catch a shark in a rowboat off the coast of Hawaii, told me to climb to the top of what scientists said was an overdue-to-blow Mount Vesuvius, and generally exposed me to mortal danger in other fun ways. He also got me to the emergency room in Chile when I needed that life-saving operation, and I owe him my life. We had been through many stories together, but in all that time, we had never talked about story itself. Now Graham was a professor at the Center, teaching science communication, so I was curious. How did he go about teaching story?

"First of all, I often show that little video that we did with the baby at Yale."

The video was an experiment we shot for a miniseries called *The Human Spark*. The star of the experiment was Nora, a curious, alert, six-month-old girl.

Karen Wynn, the Yale researcher, had set up a puppet show in which each character was simply a geometric shape with a pair of buttons for eyes. The figure was simple and cartoonish, and yet Nora was riveted by this abstraction of a face. A red disc with eyes was struggling to move up the side of a mountain, but when it got near the top, a yellow triangle with eyes would push it back down the mountain. It wasn't until a blue square with button eyes came in behind the red disc and helped by pushing it up the mountain that the red disc made it all the way to the top.

After the puppet show, when Nora was given the chance to choose which of the characters she liked, she immediately reached for the cooperative blue square. This was the choice of almost every child in the study. It was a fascinating look at how, even as infants, humans seem to have a preference for those who are cooperative, and to have a sense of right and wrong behavior.

But Graham came away from the shoot wondering if there wasn't something else also happening in the child's mind. He said, "I interpreted it, in addition, as meaning that the baby is able to put a *story* to it, is able to interpret what was going on *as a story:* 'Oh, I see. That shape is trying to get up the hill, and this other shape is trying to push it down, and this one's trying to push it up.' There's a little story that forms in the baby's mind to interpret what she or he is seeing."

This was an interesting idea, but we wondered if Graham was over-interpreting.

We got in touch with Karen Wynn, who did the study. "I don't at all think you're over-interpreting," she said. "In order to sense a meaning—good or bad—to the actions, there has to be a narrative of the kind you describe: 'This one's attempting to do such and such, that one comes along and attempts to help, but this other one comes along and tries to hinder.' It doesn't make sense without this narrative."

The important point, she feels, is that "babies are following events and understanding them in *context.*" Keeping track of the context can cause the baby to prefer the more aggressive behavior. "Sometimes," she said, "babies prefer a hinderer to a helper—when, for instance, the recipient of the hindering is someone who was mean to another guy. The specific action of helping or hindering isn't itself judged as 'bad' or 'good,' but judged in the context of what came before."

"So, in a sense," I said to Graham, "a feeling for story seems to go back all the way to infancy, doesn't it? Those kids were really young."

"Six months. They were barely able to do more than gurgle, but they were able to uniformly interpret what they saw and turn it into a little narrative in their heads. Then, all through childhood, that's what you do: 'Once upon a time . . .' It's all stories."

When Graham said the babies were able to "interpret" what they saw and turn it into a narrative, something clicked for me. I remembered a leading brain scientist I had interviewed years earlier, Mike Gazzaniga. Mike had done breakthrough work with patients who had lost the connection between the right and left hemispheres of their brains. (In order to stop epileptic seizures, surgeons had severed the cable that connected the hemispheres [their corpus callosum], leaving them effectively with two brains, each of which functioned on its own.)

Sometimes it was hard for a person with two brains to make sense of what he saw, but Mike identified a sense-making function in the left brain that he called "the interpreter."

Mike would show two pictures of objects to a patient in such a way that only his left brain could see one of them—say, a picture of a chicken claw. Mike would ask him to choose which of four other pictures the claw best went with and to point to it. With his right hand, the hand connected to his left brain, the patient would pick the picture of a chicken as the picture that went best with a claw.

The other picture, also shown at the same time but to the

right side of the brain, was a picture of a snow scene along with pictures of four different objects, including a shovel. The patient would correctly point with his left hand (the one connected to his right brain) at the shovel as the object that best went with the snow scene.

With each half brain pointing to its own answer, the patient was then asked, "Why did you do that?" One side of his brain knew the shovel went with snow and the other side knew the chicken claw went with the chicken. But in fact one side didn't know why the other had made its choice. This is when the left-brain "interpreter" got to work. When the patient was asked *why* he felt those pictures went together, his left brain had to make sense of his two choices. Again, the left-brain interpreter knew why the chicken was chosen, but in fact did not know why the left hand was choosing the shovel. Unconsciously turning it into a rational choice, the left brain deftly said, "Well, you need a shovel to clean out the chicken shed."

Mike did this experiment thousands of times and saw repeatedly that when the right and left brains, operating independently, made strange choices, the interpreter would always come up with some plausible explanation. As Mike has written: "Though the left hemisphere had no clue, it would not be satisfied to state it did not know. It would guess, prevaricate, rationalize, and look for a cause and effect, but it would always come up with an answer that fit the circumstances."

I wondered if there was some connection between this interpreter function and how we rely so much on story. I called Mike and asked him if he thought they were related in some way.

He did.

Mike said, "The vast majority of everything we do goes on outside of our consciousness, everything from touching the end of your nose to trying to think about the next sentence you're going to say. What comes out of that is how we behave or what we say. We have this left-brain monitoring that's constantly trying to make sense out of that. I'd say that this interpretation, the storytelling, is fundamental to how we express ourselves to others and how we integrate others to our space."

"You mean this literally as *story*?" I asked him. "You're not using the word *story* analogously?"

"No, no, no, it's a story. It's how we best understand and can retain the meaning of what our actions are in any given moment."

The neuroscientist Antonio Damasio makes the same point when he calls stories "a fundamental way in which the brain organizes information in a practical and memorable manner."

But after hearing from these scientists that storytelling is so much a part of our makeup, I was surprised by how hard it is for many people to understand they have a story to tell.

THE STORIES IN OUR LIVES

In his class, after Graham shows scientists the video of the infants watching the drama of the shapes, he tries to get them to see the stories in their own lives. "One of the things we try to do," he said, "is remind them that their scientific lives are full of stories, which they don't even appreciate as stories—even though every experiment is a story. In every experiment, there's a question which is raised in the scientist's mind; he or she thinks they can build on that knowledge, create something new. They come up with a plan. All sorts of things go wrong: The experiment fails, somebody quits, they drop the beaker. It's a struggle. Then, finally, there's a resolution. We've gone to a higher level. Or maybe not, maybe we haven't gotten anywhere. But even if it's failed, we've learned something."

THIS IS YOUR BRAIN ON STORY

Is story really that important in helping us communicate? And if it is, how does it do it?

The answer may lie in what story does to our brains. According to Uri Hasson, a neuroscientist at Princeton, when you tell me a story, our brains get coupled in a very real way.

Here's the experiment he's done that shows that: Someone watches a movie he's never seen before, while lying in an fMRI machine. Then he's asked to tell the story of the movie to someone else. His story is recorded, and so is his brain activity while he tells the story. Interestingly, the areas of the brain that are active while he's watching the movie are also active while he tells the story of the movie. But the real surprise is that when the recording of his telling the story is played for other people, during their own fMRI sessions, the *same* areas are active in *their* brains. It's as though they're watching the movie, even though they're only hearing the story. This activation takes place moment by moment in the story. As Uri told me, "What we see is that basically by *speaking,* you are activating the pattern of responses that are unique to the moments of watching the movie. . . . It makes you look as if you really watched the movie."

"So, it sounds like you're saying that people are more in tune with one another if they trade information through story."

"So far, we see that storytelling is amazing; it has the power to make people really aligned and more effective passing information between them."

I know, in my own experience, people have a harder time understanding what I tell them if I don't tell it as a story.

I'm at a large dinner party, sitting next to someone who, graciously fulfilling her obligation to make conversation, asks me what I've been doing lately. "Making any interesting movies?"

How do I start? I love acting, but I can get really excited by the work I do with scientists, and I like to share it with people. But for some reason, I don't tell her a story. I tell her facts. "I helped start the Center for Communicating Science and I spend most of my time helping scientists and doctors communicate better with the public and with policymakers, and in fact . . ." I don't get to the end of the sentence before I see a neon sign in her eyes flashing a distress alert: *Oy,* she's thinking. *What's all this about? How am I going to pretend to be interested in whatever this is?*

I could have told her instead about the young medical student who helped a woman actually understand her diagnosis, and even sat and cried with her. Then after my dinner partner asks me how in the world we had been able to help a young doctor learn to do that, I could tell her a couple of facts. Because now she'd want to know. As Steve Strogatz told me, the trouble with a lecture is that it answers questions that haven't been asked.

STORYTELLING AND MATH

I talked to Steve just as he was about to start a course he had never taught before in an area of math called analysis. It's a challenge for many students, he said, because it involves a shift in their thinking. Instead of just solving equations, as the students have been doing since high school, they'll now be required to come up with proofs—proofs of the algebraic process itself. And, according to Steve, students find this both boring and hard, "a double whammy." Their frustration is fueled by the thought that they have been solving equations successfully all this time and suddenly they have to start proving that the very basis of their mathematical reasoning is valid. But learning to do that will prepare them for the rigors of advanced mathematics, where they will come up with conjectures that then have to be proved.

"The way it's usually taught is you go to the end of the story and teach the end first. Because by the end, everything has been figured out. The mysteries have been solved." But the question Steve is exploring is this: Should you tell the story of trial and error that led to progress in math, or teach the math as though each new discovery logically follows what came before? In other words, should you communicate things historically or logically? The difference between these two approaches is dramatic.

As Steve described it, "The historical way is a tale of confusion and intrigue and rivalry and argument—and all that human, bloody stuff. Whereas the logical tale takes out all the blood and all the mystery, and everything is nice and organized and tidy. Mathematicians think that's the way you should teach. But I think that's exactly backwards."

The way he decided to teach the math course was, instead, to start with the human side of the story. "Who are the people who were confused in 1750 and why were they confused? And who corrected them—except those corrections were a little bit wrong, too. And then who improved on *them*?" He would boil 150 years of math down to a one-semester course. The hope was to convey, not just the mechanics of math, but the excitement of it. "I think it's a very thrilling adventure story," he said.

This is about teaching math to people who have been studying it for years. What hope is there for engaging laypeople without the human, bloody part of the story? Maybe not much.

The renowned physicist and mathematician Freeman Dyson has said, "It is impossible to write a readable book about real mathematics for nonmathematical readers. The best anybody can do is to write about a real mathematician."

It seems that our greatest hope in introducing us to the hard stuff lies in some kind of story.

THE BUILDING BLOCKS OF A STORY

So, what *is* a story? Aristotle is often quoted as saying that a story should have a beginning, a middle, and an end. That's true, but I don't think that's the whole story. After all, a dead cat has a beginning, a middle, and an end.

He meant, I think, that the middle section has to go into a different gear.

Interestingly, this is what happens in a good *60 Minutes* piece. About a third of the way through, the story deepens, or takes a turn that surprises, and you're engaged a little more fully.

Without going into second gear, a story is liable to be a simple chronological ordering of events: *This happened, and then that happened, and then this other thing happened.* It might make you wonder idly about what's going to happen next, but it doesn't grab you and deposit you into a state of suspense.

What does do that, I think, is what Aristotle came up with when he was figuring out what made *Oedipus Rex* a good play: the idea of *dramatic action.* The hero is trying to achieve something of (to him) great importance, but suddenly there are obstacles in the way. Those obstacles have to be overcome before the objective can be achieved. There's suddenly tension because everything could fall apart and end in disaster. If the audience has identified with the character, it's hard not to have an active response, like "What's he going to do about *this?*"

I guess it's because I come from the theater that I have a special fondness for *dramatic* stories. All my life, night after night, I've sensed the engagement of the audience as they followed the action and rooted for the hero to win the struggle. It didn't matter if it was a serious drama or a trivial comedy; as long as someone wanted something desperately and had obstacles thrown in his path, the audience was engaged.

I try to make this idea vivid, as I've described in another book, whenever I speak to an audience about communicating. I ask for a volunteer from the audience and ask her to carry an empty water glass across the stage. This usually evokes a few giggles, because it's hard to do without being a little self-conscious. Carrying an empty glass doesn't have much meaning.

But then I fill the glass with water—so full that if I poured even one more drop it would spill over the top. I ask her to carry the glass back across the stage and put it down on a table. "But be careful," I say. "If you spill even one drop, your entire village will die."

Everyone in the hall knows there's no village and no one's going to die. But that imaginary obstacle is powerful enough to rivet the audience's attention on the glass. If even a bead of water runs down the side of the glass, you can hear them gasp.

After an agonizingly careful walk to the table where she puts the glass down, I ask the audience, "So, which trip across the stage was more engaging?" Their laughter is the answer. We can identify with someone who has a goal, but we *root* for someone with both a goal and an obstacle.

Having a goal in the first place is crucially important. What does the hero or heroine want? What has she set out to achieve that is, at this moment, the most important thing in her life?

I don't step out on a stage as a character in a play without reminding myself of what I as a character am going out there to accomplish. It's vital to drama. I've often thought there ought to be a sign in the wings that says, "No one permitted beyond this point unless you know what you want with all your heart, and you know how you're going to get it."

It's such a useful thing to remember that I keep it in mind even when I'm going out to speak to an audience as myself: *Why am I here? Who am I talking to? What do I want to accomplish? What's my mission?*

And then I tell stories. Stories with obstacles in them.

As important as the goal is, it's not until the obstacle shows up that the story gains tension; the obstacle kicks us into second gear and raises the stakes. It's not just that we might not achieve our goal; we might lose everything. The whole village could die.

Suddenly, the middle has meaning and Aristotle's dead cat comes to life.

You can see this at work in the graphic Christine O'Connell created to teach storytelling in our writing classes.

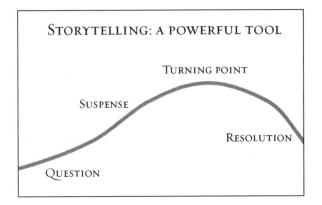

The arc of a story starts on the left of the drawing, with a "question." For Christine, this is where you let the listener in on what the leading character is trying to achieve, or, as she says, "What's the story going to be about?" Christine is herself a scientist and she sees the beginning of a story as something close to the scientific process. "The question is the hypothesis," she says. "What are you setting out to figure out?" But in an experiment, you don't always figure out in the end what you had hoped to figure out. In an experiment, in life, and in a good story, obstacles get in the way.

Here's where the importance of the middle comes in. If an obstacle complicates the story and puts everything in doubt, then we have suspense and at the height of that suspense there's going to be a turning point, where things are either going to get a lot better or a lot worse.

There's something very engaging about this because it's difficult not to be caught up in someone's struggle to achieve something. Seeing someone take a stroll on a mild day doesn't capture our attention the way watching him leaning into a gale force wind does.

THE OPPOSING THOUGHT

I may be overly enthusiastic about it, but I think this kind of oppositional force can even liven up an essay. If you propose an idea and then let it be countered by an opposing idea, it's like wind in the face.

I see it in that glorious essay Lincoln spoke: "Four score and seven years ago our fathers brought forth on this continent, a new nation, conceived in Liberty, and dedicated to the proposition that all men are created equal."

Opposing thought: "Now we are engaged in a great civil war, testing whether that nation, or any nation so conceived and so dedicated, can long endure."

We're moving forward and suddenly there's an obstacle in our path. Can we long endure? We're paying attention in a way we weren't before. Now the trip across the stage isn't so smooth; there's water in the glass that could disastrously spill. We want to see how this struggle to move forward will turn out.

In this way, I think, an essay can become a dynamic conversation with the self. It has the chance of being a little less didactic, less like the tiresome uncle at the Thanksgiving table who has the answer to his own questions and doesn't consider the possibility of opposing thoughts.

It's even broader than that.

For me, the acknowledgment of an opposing thought is one of the things that makes science such a dramatic thing to watch. The scientist says, in effect, "It looks like something is happening here—but am I wrong?" And then that opposing thought, that courageous application of doubt, takes us on an adventure of risk, tension, suspense—the emotional turmoil of experiment. And finally, we reach a turning point where, identifying with the scientist, we either achieve new understanding or we don't. And that leads to the story's resolution—a new way of seeing, a sense of meaning, that we didn't have before.

This struggle to resolve an opposing force can lift us out of the technical and plant us in the emotional, where we're engaged and root for the outcome.

An Olympic record can be decided by a fraction of an inch or the tick of a second, but the victory is far more thrilling if we know that this will be an athlete's triumph over setbacks.

And just as every scientific experiment and every scientist's life has a vivid story at its core, with built-in, inevitable setbacks, so does every business struggling to make it to the top.

GLASSBABIES: CONNECTING WITH CUSTOMERS

When I read about one relatively small business in the *New York Times*, I couldn't resist clipping it out. Lee Rhodes had a shop in Seattle that did a thriving business selling candleholders she called Glassbabies. Business was so good she decided to expand to New York City, but things didn't go so well there. Her prices seemed too high, the customers didn't drop in the way she had hoped, and when they did, the glass candleholders seemed too heavy to lug home. (Unlike in Seattle, her customers hadn't pulled up to the store in their cars.) And the high rent she had to pay for the store wasn't helping. It just wasn't working, and after a year or so, she closed the shop. When she thought about what had led to the failure in New York, she realized she hadn't planned well enough. She regretted not having done a foot-traffic study, and not having realized that the habits of New Yorkers didn't include the ability to put a sack of the heavy candleholders in the "back of their Volvos." She also realized something maybe even more important: She hadn't told her story.

What Seattle knew and New York didn't was that Lee Rhodes was a survivor of three bouts of cancer who had had an experience during her illness that she felt helped in her recovery. A small votive candle she had placed in a cup soothed her. She hired artists to design candleholders, and started giving the results to friends. Then, calling them Glassbabies, she began selling them out of her garage. She remembered the people she had met in doctors' waiting rooms, many of whom, as she says on her website, "could not afford day-to-day costs like bus fare, childcare, and groceries during chemotherapy," so she put aside 10 percent of her profits for charities that offered help to people suffering from cancer.

"It's more than a candleholder," she told the *Times,* "but we didn't do a good job explaining that and training our employees in New York to communicate that with customers."

A year later, she opened another branch store—in San Francisco. This time, she researched the neighborhood and was thousands of miles closer to her staff to train and manage them.

And, crucially, now she tells her story. On her website, she lists the dozens of projects to which she donates 10 percent of her profits.

The profits are mounting. As of this writing, she's given over three and a half million dollars to charities that offer "financial and emotional assistance to those in need." Using foot-traffic studies and being close enough geographically to train her workers was probably essential to her success. But, it seems, so was telling her story.

When a scientist needs funding, or a business needs to capture the imagination of the public, or a parent just wants to impart wisdom to a child, it seems to me that the most vivid way to do it is with a good, dramatic story. We just have to remember that when we tell the story we can't make the trip our leading character makes across the stage too easy.

It's not dramatic to carry an empty glass. We have to fill it to the brim.

CHAPTER 19

Commonality

Story may be the royal road to communication, but according to Uri Hasson, the neuroscientist at Princeton, there's a crucial component to storytelling that can't be ignored, or it will be a bumpy road.

People's minds will sync up in the presence of story, but the process relies heavily on there being some similarity between the storyteller and the listener. "The more commonalities you have with the speaker," he told me, "the better the understanding."

"What kind of commonality?" I asked. "Is it education? Is it geographical? Upbringing? Culture? What kind of commonalities are critical?"

"I think all of the above."

The more commonality between the storyteller and the listener, the more an MRI will show their brains in sync. Uri figures this is because the speaker is making use of what the listener already knows about various elements of the story, and is putting them together in a new way. But I wasn't quite sure how this worked.

Uri said, "When I'm remembering, over dinner, the movie I saw last night, I have a memory of the movie. Right?"

"Right."

"You, listening to me describing the movie, have no memory of the movie because you never watched it. Right? Let's say it was about a detective in a murder story and it's happening in New York City. I need to activate my memory [of the movie], and you need to take your knowledge of what a detective is, and a murder, and driving in a cab in New York City, in order to understand what I'm saying. Right? There is no memory of the movie, but still you're understanding it and it looks like you've watched the movie. Right?"

"So, in a way," I said, "you're helping me put together parts of my memories, and creating a context for me for the story and the things that happen in it."

"Exactly. If you have no clue what a murder is, you're completely lost. Right? There's something interesting about communication. Communication works only in cases when you understand something about what I'm going to say to you."

"Which is probably why teachers make sure they know what the student already knows so they can build on that."

"Yeah, exactly."

But there's a catch. For the best communication to take place, it may be that we can't just be alike; we may have to be *aware* we're alike.

MAKING US AWARE OF OUR FAMILIARITY

Jessica Lahey is a writer and a teacher. She told me she became aware of how important familiarity can be when she was denied it.

She was teaching writing at an inpatient drug and alcohol rehab center and was frustrated because she wasn't getting through to her students. She wasn't sure why, but she felt as if she were teaching in a "hermetically sealed bubble." She was on the inside and they were on the outside.

Her frustration mounted until, by chance, she met someone at a conference who could analyze the problem. Hunter Gehlbach had done a study on the effect of familiarity in teaching, and after speaking with him she realized why she had felt hermetically sealed.

Unlike with her other teaching experiences, at the rehab center she had to adhere to strict rules on confidentiality that kept her from knowing anything about her students' lives and kept them from knowing anything about hers. She and her students had no way of knowing if they shared any points of commonality.

Gehlbach told her that "figuring out the thoughts and feelings of others seems key." And that teachers need to be able to "figure out the thought processes of students as much as possible, in order to understand where and why they're making mistakes."

He looked at learning as a *social* process, an interaction, and explored the *relationships* between students and teachers.

Gehlbach studied hundreds of ninth-grade students and their teachers, giving both the students and teachers a questionnaire at the beginning of the school year. He asked them to list their preferences about things in their lives—like hobbies or what they thought characterized a good friend. After five weeks, he gave the teachers and students another questionnaire and found that those teachers who saw that there were five points of commonality with students tended to regard those students as more familiar—and to feel they had a better relationship with them than with others. And that feeling of commonality had an impact on learning. As a result, Gehlbach says, "those randomly selected students earned higher grades in the class."

All this from five points of similarity.

As Gehlbach pointed out, this is only one study and it would be premature to make broad recommendations. But even though any single study is seldom the final answer to anything, it can point you in a direction that's worth exploring.

And an awareness of similarity seems to be worth exploring, because it's helpful in figuring out what the other person is thinking.

Scientists at Harvard, for instance, did MRI studies on people while they were attempting to read the state of mind of others and found they were better able to read their state of mind when they saw themselves as similar to the other people.

A sense of similarity seems to help us sync up with one another. If that's true, it probably doesn't hurt to remind the people we're communicating with of the points we have in common: The musical tastes a mother shares with a daughter, along with a few other points in common, might ease an otherwise more difficult conversation.

For me, the most striking example of the power of similarity is the story of the so-called Christmas truce in 1914, during World War I.

When Great Britain entered the war, many of the troops were sure they'd be home by Christmas. But as months went by and Christmas approached, they were still stuck in the trenches, with icy puddles, rodents and lice, and mold in their bread. After a while, the British noticed that the other side stopped firing around dinnertime—and then, an hour later, they would resume. The British took advantage of the lull to have their own meal. A routine evolved: As if to emphasize that they were taking a timeout from the war, both sides would fire furiously right up until the clock struck five and then there would be utter silence. At precisely six o'clock, the guns would begin again.

Soon each side began making occasional offerings of a break in hostilities whether or not it was dinnertime. At one point, in a clear appeal to familiarity, there was a crucial bit of communication. A German voice called out, "We are Saxons, you are Anglo-Saxons. If you don't fire, we won't fire." By Christmas Eve, the similarities in the culture they shared were becoming hard to ignore. The British could hear the Germans singing Christmas carols.

Malcolm Brown, of the British War Museum, quotes a soldier remembering what he heard from his position in a trench in La Chapelle-d'Armentières: "It was a beautiful moonlit night, frost on the ground, white almost everywhere. And about seven or eight in the evening there was a lot of commotion in the German trenches and there were these lights." The lights turned out to be candles burning on improvised Christmas trees. "And then," the soldier recalled, "they sang 'Silent Night'—*Stille Nacht*. I shall never forget it. It was one of the highlights of my life. I thought, *what a beautiful tune.*"

The British soldiers answered the Germans by singing "The First Noël" and held up signs on which they had scrawled, "Have a happy Christmas." They sang through the night, matching each other song for song.

On Christmas morning, in many places along the front lines, men from both sides slowly climbed to the tops of their trenches and met in the middle of no-man's-land, where they exchanged cigarettes and offered one another schnapps and chocolates. In three or four places along the line, someone was able to come up with a ball and, placing their hats on the frozen ground to mark off a pair of goals, the two enemies played soccer.

One British infantryman even had his hair cut by a man who had been his barber in London and was now a German soldier. There was a sense of relief and a measure of hilarity for both sides.

I tell this story because it seems like such a profound example of the power of commonality. If people are shooting at you repeatedly for months, and if reminding them you share something in common can silence the guns for a while, something important is going on. It ought to be even easier to do when all you want is to get them to pay attention to what you're saying.

There were, however, a few generals on both sides of the war with whom all this familiarity did not go over so well. They let the troops know that fraternization was treason, and from then on the only time either side would climb out of the trenches was with fixed bayonets.

In order to fight a war (probably the most extreme form of poor communication), *un*familiarity is the preferred state of mind.

LETHAL UNFAMILIARITY

By the time U.S. forces entered the war in 1918, General John J. Pershing felt that the French soldiers had become passive and less aggressive, and he was worried that trench warfare would affect his American troops in the same way—especially because they seemed reluctant to stab other humans in the neck.

He made sure the Americans practiced on bayonet dummies filled with straw. The object was to rid the enemy of any sense of familiarity and dehumanize the target of your bayonet. You were to aim at a circular patch on an inanimate bag, and if you thought of the opponent at all, it was as a monster. One instructor was quoted as saying, "When you drive your bayonets into those dummies out there, think of them as representing the enemy. Think that he has begun the practice in this war of running bayonets through wounds, [those] gasping-on-the-ground and the defenseless prisoners. . . . So, abandon all ideas of fighting them in the sportsmanlike way. You've got to hate them."

While in war we doggedly emphasize our unfamiliarity with the enemy, in peace we often reach awkwardly for signs of how alike we are.

Campaigning politicians go through bizarre rituals to fit in with the locals, like eating deep-fried things on a stick when in fact they might prefer a little quiche and white wine. The rituals themselves are so familiar to us we don't even question how fake they might be. They're just what people do to get elected.

Pulled by the magnetic field of the familiar, we go to sequels of movies, even though we suspect they won't thrill us the way the originals did.

Some studies have even shown that romantic attraction is more likely to occur when people show up repeatedly in your daily life. Which, I suppose, is the case with the office romance.

I think all this points to the idea that as long as it doesn't seem fake, the more we establish familiarity with our audience— not speaking to them from left field or from on high—the better chance we have that they'll listen to what we have to say. And possibly even accept it.

CHAPTER 20

Jargon and the Curse of Knowledge

THE IRRESISTIBLE AROMA OF HYDROFLOXIA

My wife and I are walking in a garden. It's spring. New life is curling up toward the sun from the black, wintry earth. As we walk, she names the flowers beside the path. Hyacinth, ranunculus, iris. She names them because calling them by name is for her part of loving them. Happily, she goes on—anemone, crocus, lily of the valley. After a while, I can't stand it anymore. I point to one she's missed.

"Look at that gorgeous hydrofloxia," I say, and immediately I feel a surge of pleasure at having inside knowledge. Arlene is not impressed. She knows I'm making it up.

For one brief moment, I had enjoyed speaking the private language of botany. It didn't bother me that the word doesn't exist. We both knew I was joking, but I got to use a fancy word and I loved it. There's something appealing about a private language. It can be intoxicating. Jargon is like that, and the more rarefied it is—the fewer people who understand it besides you—the more it resembles the common hydrofloxia. It has a seductive aroma. You can get drunk on it.

A nice example of this is the research paper by Ike Antkare called "Developing the Location-Identity Split Using Scalable Modalities." The title is impressive. And so is the first sentence of the introduction:

> The implications of atomic communication have been far-reaching and pervasive. The notion that steganographers connect with "smart" archetypes is continuously considered intuitive. Along these same lines, this is a direct result of the development of the World Wide Web. Thus, the investigation of write-back caches and DHCP have paved the way for the refinement of e-business.

The only problem with this paper is that, from beginning to end, it's utter nonsense. I don't mean it *seems* like nonsense; I mean that it's deliberately crafted to mean nothing, and it wasn't even written by a human. The author is a machine.

Ike Antkare is the name of a fake scientist created by the French researcher Cyril Labbé, who wanted to see if a mere golem like Antkare could make a name for himself on the Internet among real scientists.

He certainly could. Labbé used a computer program called SCIgen (created by graduate students at MIT) to generate dozens of fake papers using just a few keystrokes. In a short time, the fictional Ike Antkare had so many citations, he became, according to Labbé, "one of the great stars in the scientific firmament." This happened mainly because the fake papers were picked up by Google Scholar, which is used by other organizations to rate authors on how many times they've been cited.

After Labbé was finished gaming the system, Google Scholar ranked Ike Antkare number twenty-one on the list of most cited authors. Ike didn't have as many citations as Sigmund Freud, but he had more than Albert Einstein.

The golem ruled.

Labbé went further. He developed a program that could detect whether a paper had been created using SCIgen. He found 120 papers that had been accepted and published by peer-reviewed journals—all created by SCIgen and all fakes. The journals pulled the papers after Labbé notified them, and at least one said they would revise their acceptance process.

I think it's interesting that it wasn't unintelligibility that caused the journals to pull the papers, but *fake* unintelligibility.

The three computer scientists at MIT who created SCIgen had shown how easily jargon gone amok can open the door to fraudulent research papers: nonsense leading to non-science. They invented SCIgen as a way for other scientists to test conferences for their reliability. For instance, how easy would it be to get a bogus paper past them? It turned out to be surprisingly easy in too many cases.

There are other jargon generators online. I used the "Automatic SBIR Proposal Generator" to write a fake application for a grant from the National Institutes of Health (NIH). It took less than a minute to create, and the full thing goes on for several pages. Here's just a bit of it.

The Hydrofloxia Effect
Technical Abstract

The technology in The Hydrofloxia Effect effectively addresses the indirect paradigm causing the bandlimited criterion by applying a laser-aligned orthogonality that varies. This technology will provide NIH with the quantitative memory. . . . The successful development of The Hydrofloxia Effect will result in numerous spinoffs onto a symmetric mainframe for the benefit of all people in the world.

It's heartening to know that the study of hydrofloxia will benefit the people of the world. Even so, I decided not to submit my breakthrough research paper to the NIH. They're too busy deciphering real applications.

GOOD JARGON

There are actually some nice things to say about jargon. First, of course, we have to recognize that there probably isn't a line of work that hasn't developed its own jargon. If you walked onto a movie set and someone asked you to "go get the gobo on the Century over there, and while you're at it bring back a half apple and a kook—and hurry up, this is the Martini Shot," you might be a little puzzled. Among other things, you're being asked to get a couple of things that cast shadows. The gobo casts a hard shadow and is attached to a Century stand, manufactured by the Century Company and bearing its name. The kook, or cucoloris, is a board with a patterned cut-out for casting feathery shadows. A half apple is a small platform about the size of half an apple box. Cameras, lights, and height-challenged actors can be placed on them. And the Martini Shot is the last shot of the day, after which everyone goes home and has a martini.

The point of running through all this arcane etymology is that, although jargon often proceeds from misty origins, it usually has a specific and useful meaning. Sometimes one word can stand for five pages in plain English. If people in the same field share a knowledge of that meaning, they're not going to use five pages if one word will do, and they shouldn't be expected to. Speaking jargon to the right person can save time and it can also lead to fewer errors. "Bring me the gobo" is probably less prone to error than "Bring me the black fuzzy thing over there."

But the other person does need to define the jargon in the same way that you do. I heard about a meeting in Washington where a group of nanoscientists were brought together with a group of neuroscientists in the hope that they could collaborate on new ways to study the brain. Before they could even get started, they wasted hours in a cloud of confusion because they couldn't agree on the meaning of one word: the word *probe*.

Even an ordinary English word can become indecipherable jargon if the person you're talking to uses it as their own indecipherable jargon.

There's another reason jargon creeps into almost every profession: It can feel good to speak jargon. It's like the feeling I had the day I came up with *hydrofloxia*. I had the pleasure of speaking a secret language. It was a little too secret, because nobody else speaks it, but a private vocabulary is not a totally bad thing. People seem to bond when they share shorthand. *We're the ones who talk like this. At least in this way, we're special.* It may seem like a trivial way to bond, but if it helps build a team, it probably doesn't do harm—*unless* it's used to keep people out who ought to be let in. A doctor speaking to a patient in terms she can't understand is certainly not engaged in a bonding experience.

Probably the least useful reason for using jargon is that it makes us sound smart. If the other person understands the jargon, then he or she is just as smart as the person speaking, so there's nothing gained. And if the listener does not understand jargon, then it probably doesn't sound all that smart to be unintelligible.

Jargon is dangerous because it usually buries the very thing you most want the other person to understand. The insidious thing about jargon is that we know how beautifully it expresses precisely what we want to say, and it simply doesn't occur to us that the person we're talking to doesn't have a clue as to what we're talking about. In these moments, we suffer from a very common malady.

THE CURSE OF KNOWLEDGE

I first became aware of this strange mental lapse when I read *Made to Stick,* the very useful book on communication by Chip Heath and Dan Heath. Steven Pinker also talks about the curse of knowledge in his book on style. A number of writers have latched onto the concept as being central to good communication. I agree. I think it's at the very heart of communication. But, interestingly, the term originated in 1989 when three economists wrote a paper, not on communication, but on business and finance.

Colin Camerer, George Loewenstein, and Martin Weber did a study to see if it's true that having more knowledge than another person, for instance about something you're buying or selling, really does give you an advantage. Their shocking conclusion was that very often extra knowledge is a disadvantage. At first it seems nonsensical that knowledge could be a burden, and even a curse. The problem, of course, is not in the knowledge itself. The problem is when you can't imagine what it's like not to have that knowledge. This is because people are, according to the economists, "unable to ignore the additional information they possess." There's something about having knowledge that makes it difficult to take the beginner's view, to be able to think the way you did before you had that knowledge. And unless you're aware that you actually know something the other person doesn't know, you can be at a disadvantage. When you forget you know more than *they* do, there's a tendency to under-value your position.

It goes something like this: You're selling a used car and you know the mileage is low, but you also know it's been beat up its whole life on bad roads. You're not required to reveal that infor-mation, but you can't forget it. You unconsciously assume the buyer will know what you know, and you set your initial price lower than you need to.

It's not uncommon for this to happen in a negotiation, or even in buying and selling stocks—whenever the more knowl-edgeable person just can't shake the feeling that if they know something, then it surely must be known by everybody.

It strikes me that this is curiously similar to how a four-year-old thinks before she realizes that other people have thoughts that are different from hers; yet, this is a flaw in *adult* thinking. Accomplished adults, at that.

Heath and Heath use an example from the world of business, in which an executive attempts to rally the workforce with an abstract strategy like "unlocking shareholder value." The execu-tive has condensed a lifetime of concrete experience into an ab-stract term that to a workforce lacking his experience sounds like a foreign language.

Once we know something, it's hard to unknow it, to remem-ber what it's like to be a beginner. It keeps us from considering the listener.

Using shorthand that is incomprehensible to the other per-son, or referring to a process they're unfamiliar with, we lock them out, and we don't even realize it because we can't believe we are the only person who knows this stuff.

It was through Heath and Heath that I learned about a re-markable demonstration of the curse of knowledge, invented by a graduate student at Stanford in 1990. I've borrowed it and shown a version of it to audiences, who, almost without fail, get the idea instantly.

The grad student, Elizabeth Newton, worked out a simple experiment at Stanford in which she divided people into two groups: *tappers* and *listeners*. The tappers had to decide on a well-known song, like "Happy Birthday," and let the listener know what the song was. But they couldn't hum the tune or recite the words; the only way they could communicate the song was by tapping out the rhythm on a table. Take a second now and try to guess how often the listeners were able to identify the song. Eighty percent of the time? Fifty percent?

According to Heath and Heath, out of the 120 songs that were tapped out, the listeners were able to guess only 3 correctly. In other words, they succeeded only 2.5 percent of the time. How does that match with your estimate?

When Elizabeth Newton asked the tappers what percentage of the listeners would be able to figure out the songs, their average estimate was about 50 percent, and this is what I've found with audiences. I've done the game, by now, with a few thousand people, and on average the tappers think that about half will identify the song, with some estimates going as high as 80 percent. But almost always only 2 or 3 percent of the audience can recognize it.

The problem for the tapper is the peculiar disadvantage of the curse of knowledge. It's almost impossible to tap out the rhythm of a song without hearing the melody in your head. Once you hear the melody, there's a strong tendency to assume at an unconscious level that the people listening hear it, too.

It happens when people playing charades repeat over and over the same gesture, certain that it conveys the word they hear in their head.

This is the irrational notion, first examined by the three economists, that other people know what you know.

When a scientist uses language that's just beyond the audience's reach, or when a doctor describes a medical procedure in terms the average patient doesn't understand, the scientist and the doctor hear the melody but the people listening only hear the tapping, and they're liable to think it's another song entirely.

To me, the melody of nature is a beautiful song, and if all a scientist gives me is the tapped-out rhythm, but keeps the melody to himself, he's barely given me the skeleton of nature. I want her energy, her luminous skin, the light in her eyes. I want to see nature snap her fingers and dance.

CHAPTER 21

The Improvisation
of Daily Life

There's no failing in improvisation. In improv, what some people call failure is just the next step on the way to an interesting resolution.

It's like the line from a poem by Samuel Beckett: "Try again. Fail again. Fail better." Those words rest on my desk and remind me almost every day that the ashes of failure can be a place of despair or they can be where the phoenix comes from.

Scientists *fail better* when they're looking for *more* truth rather than some absolute true-for-all-time truth. And the rest of us *fail better* when we give ourselves over to the improvisation of daily life. Things change; we accept that and go with it. Connection happens between us and suddenly we see things about one another we'd never noticed before; just as in an improv, invisible objects become real, and then they transform.

It's always possible to see more deeply into what we thought we knew, or to step back and see things through a larger frame of reference.

If we remember that every conversation we have, every bit of advice we give, every letter we write, can be an exchange in which the other person might actually have a better way of looking at it, then we have a chance to be in sync, to be in a dance with a partner. Not a wrestling match with an opponent.

But it's a dance we learn by trusting ourselves to take the leap, not by mechanically following a set of rules.

IT'S NOT A FORMULA

In my childhood, they used to print diagrams that showed you where to put your feet when you did the fox trot. *Left foot here, then it moves over here, and the right foot goes here.* It might have helped you put your feet in the right place, but I don't think it ever made anyone *dance*.

In the dance of communication, we move together with another person gracefully, pleasurably, sharing the pure animal joy of community.

Not being able to communicate is the Siberia of everyday life—a place that, crazily, we often send *ourselves* to.

But the solution, in my view, isn't a formula, a list of tips, or a chart that shows where to put your feet. Instead, it's transforming yourself—like going to the gym—only a whole lot more fun.

Practicing contact with other people feels good. It's not like lifting weights. It feels good while you're doing it, not just after you stop.

When it clicks, when you're in sync with someone, even for the briefest moment, it feels like the pleasure of reconciliation. We're no longer apart. We have an actual two-way conversation. We go from "No, you're wrong" to "Oh. Maybe you're right." And *boom*. Dopamine.

It's a good feeling. I think we crave it.

Liz Bass and I talked about this after she stepped down as the first director of the Center for Communicating Science. I asked her if, while she ran the Center, anything had been surprising to her.

"One thing that struck me," she said, "is how sort of universal this desire to communicate really is and how good people feel when they sense that connection."

I wondered why that good feeling doesn't lead to better communication all by itself. Why isn't it self-reinforcing?

Liz said, "Sometimes I think it's that people don't identify the thing that made that communication good. I think people don't pay that much attention to what they pay attention to. You know what I mean? Sometimes I tell students, 'Pay attention to what you pay attention to.'"

I agree, and I hope they'll pay attention not so much to the mechanical things, like a sudden change of pace in a talk or a sudden change in volume of their voice. I hope they'll pay attention, instead, to the fundamental *source* of that pacing and volume, which is the connection with the other person. That connection makes us respond like a leaf in the breeze to whatever is happening in the faces of those in front of us.

So, it's really not that complicated: If you read my face, you'll see if I understand you. Improv games, and even exercises on your own, can bring you in touch with the inner life of another person—even when you sit by yourself and write.

But it's an art. It's not a formula. Great things can happen—but you have to be able to read the other person. I know from experience.

Our family was on vacation in the Virgin Islands and I went on a walk with my six-year-old grandson Matteo. The air was moist and tropical. We were in a paradise of green leaves and blue sky. A light breeze cheered our skin and carried a salty whiff of ocean to our nostrils. It was almost perfect.

And then, at the side of the path, we saw a tree we'd never seen before. It had spiky thorns that climbed up along its slender trunk. It looked like a dragon's back.

Matteo pointed at it. "Look at that tree," he said. "How did it get like that?"

Now the day was perfect. He was asking a deep question. He wanted to *know*. This was a chance to introduce him to the idea of evolution.

We sat on the ground and talked about natural selection, adaptation—the whole thing. For forty-five minutes. It was glorious.

The next day, he was swimming with his cousin and asked *her* a question, and she said, "That sounds like a science question. Why don't you ask Grandpa?"

He said, "I'M NOT MAKIN' *THAT* MISTAKE AGAIN."

ACKNOWLEDGMENTS

The only sad part about writing acknowledgments is that there's going to be somebody I'll forget to acknowledge. Forgive me. You were invaluable.

I *can* remember Kate Medina, who is as sharp and encouraging as an editor can be, and I'm grateful for the gentle nudge she always gave at just the right time.

And I'm grateful, as well, to Amanda Urban, my book agent, for her professional skills and for her friendship. It's a wonderful thing to share a friendship with your agent.

And heartfelt thanks to my acting agent, the legendary Toni Howard, who helps keep my life creative and interesting. Hugs to Toni.

Deep, deep gratitude to Jean Chemay, my assistant, who is warm and thoughtful and makes sure my life happens in an orderly sequence of events. Without her, I would be roaming the streets, asking people if they know how I can get to St. Louis.

My wife, Arlene, reads every draft of everything I write and manages to be totally honest and totally supportive—both at the same time. As she is in life. She always points me toward my better self.

Thank you to the entire staff of the Alda Center for Communicating Science. Your creativity, your selfless hard work, and your determination to help the world be a better place is astounding. You inspire one another, and you inspire me.

And thank you to Laura Lindenfeld, the new director of the Center, for leading it with exciting ideas, boundless energy, and wisdom. I know that before she reads a book, she likes to read the acknowledgments. Hi, Laura.

And to all the scientists who generously let me grill them about their work, it was fascinating to talk with every one of you. I'm grateful that you've volunteered your lives to be the curious explorers of our species.

And if you've actually read this far, thank *you*, too.

—*Alan Alda*

INDEX

ABOUT THE AUTHOR

ALAN ALDA, seven-time Emmy Award winner, played Hawkeye Pierce in the classic TV series *M*A*S*H* and wrote many of its episodes. He has appeared in continuing roles on *ER, The West Wing, The Blacklist,* and Louis C.K.'s *Horace and Pete.*

He has starred in, written, and directed many films, and was nominated for an Academy Award for his role in *The Aviator.*

His interest in science led to his hosting the award-winning PBS series *Scientific American Frontiers* for eleven years, on which he interviewed hundreds of scientists. Also on PBS, he hosted *The Human Spark,* winning the 2010 Kavli Science Journalism Award, and *Brains on Trial,* in 2013. On Broadway, he appeared as the physicist Richard Feynman in the play *QED.* He is the author of the play *Radiance: The Passion of Marie Curie.*

Awards for his work in communicating science include, among others: the National Academy of Sciences Public Welfare Medal, the National Science Board's Public Service Award, the Scientific American Lifetime Achievement Award, and the American Chemical Society Award for Public Service.

In 2014, he was named a fellow of the American Physical Society for his work in helping scientists improve their communication skills. He is a member of the board of the World Science Festival and is a Visiting Professor at Stony Brook University's Alan Alda Center for Communicating Science.

alanalda.com
Facebook: @AlanAldaFanPage
Twitter: @alanalda

ABOUT THE TYPE

This book was set in Sabon, a typeface designed by the well-known German typographer Jan Tschichold (1902–74). Sabon's design is based upon the original letter forms of sixteenth-century French type designer Claude Garamond and was created specifically to be used for three sources: foundry type for hand composition, Linotype, and Monotype. Tschichold named his typeface for the famous Frankfurt typefounder Jacques Sabon (c. 1520–80).